Heat and Concentration Waves

ANALYSIS AND APPLICATIONS

Heat and Concentration Waves

ANALYSIS AND APPLICATIONS

G. Alan Turner

Department of Chemical Engineering
University of Waterloo
Waterloo, Ontario

ACADEMIC PRESS New York and London 1972

ACADEMIC PRESS, INC.
111 Fifth Avenue, New York, New York 10003

United Kingdom Edition published by
ACADEMIC PRESS, INC. (LONDON) LTD.
24/28 Oval Road, London NW1

LIBRARY OF CONGRESS CATALOG CARD NUMBER: 76-182602

PRINTED IN THE UNITED STATES OF AMERICA

To Joyce and all my family

Far be it from me to pretend that the following pages achieve what first burned in my mind with pure lambency ten years ago.

—Stella Gibbons: *Cold Comfort Farm*

Contents

2. Sources in One-Dimensional Flow

3. The Ideal Impulse in One-Dimensional Flow

4. The General Pulse; Properties and Uses of Its Transforms

5. Behavior at a Longitudinal Discontinuity: Pulses

6. Sinusoidal Waves

Preface

A wave is an aberration, a lack of tranquility, that travels through space and time. The medium that transmits it may be material or ethereal, stationary or moving. The disturbance may be single or one of a train; it can manifest itself in many ways—one thinks of a movement, a force, a strain, for example. Something other than the medium affects the wave, to generate it and detect its presence, and so, by and large, such interaction is a focus of interest.

In this book both single disturbances and continuous trains of waves are considered, but the entity that changes is either temperature or mass (as expressed by concentration); the medium is material and flowing, but among the possible influences on the wave one of the strongest is its interaction with stationary matter washed by the fluid. In brief, this book describes the behavior of a limited class of waves of temperature or concentration that travel in a continuous medium which itself is moving.

To be of much use an account of their behavior has to be mathematical and so this book could also be said to be a discussion of waves as solutions to linear differential equations; for this reason the equations are called *wave equations*. The ones considered are few, relatively simple, and chosen to have practical applications in certain fields. It will be observed, however, that they are the same as the equations describing some processes of mass or heat transference in one-dimensional flow, and which in spite of—or because of—their relative simplicity have been used to devise methods of measuring a variety of quantities. This, then, is their interest in the fields of:

Physical chemistry, chemical engineering, and process metallurgy (surface and bulk diffusion in solids and fluids, phase equilibrium, catalysis, chromatographic processes, thermal conduction, mass and heat transfer, dispersion, laminar boundary layers, structure of packed beds).

Anatomy and physiology of humans, animals, and plants (distributions of diameters of veins, capillaries, tracheae, and bronchial tubes; diffusion and dispersion; transport processes in organs).

Hydrology and petroleum reservoir engineering (structure of porous rocks).

Soil science (equilibria and kinetics of the storage and uptake of fertilizer nutrients; structure of pores in soils).

The processes to which these equations are applied are basically complicated and random in nature. It follows that any quantity determined will have been simplified in makeup and averaged in time or space; this treatment, however, is not necessarily too severely restrictive; in practice the averaged quantities are still usable in design or investigation.

The reasons for my treating time-varying temperatures or concentrations as *waves* are: to unify the analyses; to show the processes to be mutually analogous; to endeavor to reduce the number of occasions that a problem is re-solved because it appeared to be unique; to reduce the consequent plethora of jargon; and to utilize some of the well-understood attributes of wave motion that are seldom applied to physicochemical systems. Among the phenomena and appellations associated with waves are: transmission, attenuation (or absorption), reflection, refraction, diffraction, interference, coherence, scattering, impedance, admittance, reciprocity, beats, and standing waves. Many of these can be identified in the behavior of time-varying concentrations and temperatures, as Chapters 1 and 5–7, and Appendix 10 demonstrate; with a little more effort, possibly all of them could be so recognized. That is, there is little forcing needed to fit this time-varying behavior into the mold of wave motion.

As the first step along the path of unification I have used the same equations to describe both mass transport and heat transfer cases. To this end the variable v, in various fonts, indicates *concentration*, that is, the amount per unit volume of either material (dissolved or mixed), or of heat. (This concentration of heat is given of course by the product of the volumetric specific heat and the temperature.) The outcome is an immediate economy of words and mathematics, with some interesting correspondence between the two cases when an interface is involved. It seemed to me that the time was not yet ripe for using one of the more radical systems of unification that have been proposed (and which have disadvantages). However, with the system adopted the problem arises of finding single words which can be used for the corresponding attributes of both mass transport and heat transfer processes. Hence, occasionally I have used the word "solute" to mean both dissolved (or mixed) material and heat, when it was desired to distinguish these from the fluid that transported them. Furthermore, overcoming my repugnance of jargon, I have invented a word "mal" to indicate a unit either of mass (e.g., mol) or of heat (e.g., cal) to be used as appropriate. I hope that the succinctness of the text will be considered justification for taking these liberties with the language—the plea, of course, used by all jargonists. Generally, however, cgs units are used.

The sequence of the text is as follows. Chapter 1 gives an account of plane waves and pulses that deliberately avoids specific examples in order to stress its generality. The rest of the book largely treats of concentration waves in a moving medium. Thus, Chapter 2 shows how linear and planar impulses can be built up from the concept of an instantaneous point source, whereas, because of the wide interest in the planar instantaneous source (or impulse, or Dirac δ-function) Chapter 3 discusses it at some length.

Chapter 4 attacks the problem of getting quantitative information from any shape of humped pulse and introduces the complication of having stationary material affecting the wave. Chapter 5 carries forward and broadens the treatment of general pulses by dealing with a system of finite length (as real ones must be), discoursing on boundary conditions in so doing.

Sine waves, because of their many advantages, are studied in some detail in Chapter 6; here again both finite and infinite systems are covered and important wave attributes are exhibited. The effect of stationary material interacting with longitudinal waves (particularly sine waves) is enlarged upon in Chapter 7.

It is intended that the book be both laboratory handbook and treatise; accordingly Chapter 8 reviews some practical problems of measurement while Chapter 9 is an outline of computational processes. Stress has been laid on the estimation of experimental errors because of their effect—occasionally striking—on the reliability of these determinations.

Finally, ten appendixes amplify a number of topics: the transformation of variables, the evaluation of important integrals, the normal distribution curve, aspects of the Laplace transform, some forms of transport equation common to both heat and mass transfer processes, and the interference of waves.

Acknowledgments

I wish to thank my colleagues for their interest and encouragement, and especially Drs. K. S. Chang, T. Z. Fahidy, M. J. Goss, and L. Otten for their discussions. My thanks are due to Mrs. J. Spowart for her skill, accuracy, and speed in converting my egregious manuscript into typescript; to Mr. R. Koopmans for making the drawings; and to the University of Waterloo for the award of a University Research Grant to defray the cost of preparing the manuscript. Any figures that have been reproduced have acknowledgments in the legends.

As regards inspiration, just about all the quoted references have provided that; others too, for I have listed only those that introduce or amplify a topic. In particular, Moore's "*Traveling Wave Engineering*" first brought home to me the breadth of the subject of traveling waves. But the person who first aroused my interest in their behavior and potential use is James A. Storrow; hence I reserve my thanks to him to the last.

Nomenclature

(Note: "mal" means "moles or calories"; in equations [ ] indicates dimensions, and R after an equation number means the equation is being reproduced from elsewhere in the book.)

A	A constant; also scalar semiamplitude; Chapter 1
A	$= \int_0^T v(t) \cos \omega t \, dt$; Chapter 4
A	A constant; Chapters 5, 6
A	Area, cm^2; Chapters 2, 5
A	Defined by Equation (7.32), dimensionless; Chapter 7
A	Generalized semiamplitude, scalar; Chapter 8
$\tilde{A}$	Constant of proportionality; Chapter 3
$A(x)$	Defined by Equation (7.46), dimensionless; Chapter 7
A_a, A_b	Cross-sectional areas; Chapter 6
$A(s, 0)$	A constant [Laplace transform of $v(t, 0)$]; Chapter 4
a	Thermal diffusivity ($\mathbf{k}/F$) or diffusion coefficient, $cm^2\ sec^{-1}$; Chapters 1, 3, 7, 9
a	Surface area of heater, cm^2; Chapter 8
a	A constant; Appendix A3
a	Surface area of thermometer, cm^2; Chapter 8
a_{jr}	Equation (5.57a); Chapter 5

a'_{jr}	Equation (5.57b); Chapter 5
a_{mj}	Equation (5.77b); Chapter 5
$\bar{a}_{mj}$	Equation (5.77c); Chapter 5
$\tilde{a}_j$	Area of interface (phases 1 and j) per unit volume of flowing phase, cm^2 cm^{-3}; Chapters 4, 7
a_2	Defined by Equation (7.37), dimensionless; Chapters 7, 9
B	$= \int_0^T v(t) \sin \omega t \, dt$; Chapter 4
B	A constant; Chapters 1, 5, 6
B	Defined by Equation (7.33), dimensionless; Chapter 7
$B(x)$	Defined by Equation (7.47), dimensionless; Chapter 7
b	Standing value as multiple or fraction of I, dimensionless; Chapter 8
b	A constant; Appendix A3
b	Defined by Equation (6.23), cm^{-1}; Chapter 6, Appendix A9
b^*	As b, for j stationary sources of flux, Equation (9.5); Chapter 9
b_{jr}	Equation (5.57c); Chapter 5
b'_{jr}	Equation (5.57d); Chapter 5
b_{mj}	Equation (5.77d); Chapter 5
$\bar{b}_{mj}$	Equation (5.77e); Chapter 5
C	A constant, cm^3 sec^{-1} mol^{-1}; Chapter 4
C	Heat capacity, cal g^{-1} °C^{-1}; Chapters 7, 9
C	A function of z and t, Appendixes A1, A2
C	Capacitance per unit length, F cm^{-1}; Chapter 7
$\mathrm{Ci}(N_f)$	Cosine integral of N_f; Appendix A10
C_l	Defined by Equation (7.48), dimensionless; Chapter 7
c	A constant = velocity of wave, cm sec^{-1}; Chapter 1
c	Quantity proportional to concentration; Chapter 3
D	Longitudinal dispersion coefficient, cm^2 sec^{-1}
D_l	Defined by Equation (7.49), dimensionless; Chapter 7
d	Distance measured upstream from a boundary, cm; Chapter 6
d	Diameter of wire, cm; Chapter 8
E	Particular integral, Equation (5.54); Chapter 5
E	Emf, V; Chapter 7
E	Emf per degree, V deg^{-1}; Chapter 8
E'	Emf due to rate of change of temperature, V sec deg^{-1}; Chapter 8
e	Exponential
F[function(t)]	Fourier transform of function(t); Chapter 4
F	Flow rate, cm^3 sec^{-1}; Chapter 9
F	Function
F	Volumetric heat capacity, cal cm^{-3} deg^{-1}; Chapters 1, 6, 7, Appendix A8
$F(s)$	Equation (4.33); ratio of transformed concentration at two stations; Chapters 4, 9
f	Function, Chapter 5, Appendixes A1, A2, A3
f	Frequency, cycles sec^{-1}, or revolutions sec^{-1}; Chapter 1
f_1	Function of parameters; defined by Equation (9.16); Chapter 9
f_3	Function of parameters; defined by Equation (9.23); Chapter 9
G	Conductance per unit length, mhos cm^{-1}; Chapters 1, 7
G_E	Voltage gain in Figure 8.9, volts output/volts input; Chapter 8

G_M	Gain of multiplier, volts output/watts input; Chapter 8
$G(i\omega)$	Response = ratio of Fourier transforms of output to input, Chapter 4
g	Defined by Equation (6.25), cm^{-1}; Chapter 6, Appendix A9
g^*	As g, for j stationary parallel sources of flux; Equation (9.6); Chapter 9
H	Defined by Equations (7.16) and (7.17), dimensionless; Chapter 7
H	Heaviside unit function; Chapters 5, 8
H	Specific enthalpy, cal g^{-1}; Appendix A8
h	Heat transfer coefficient, cal sec^{-1} cm^{-2} °C^{-1}; Chapters 6, 7, 9, Appendix A8
I	Denotes integral; Chapter 2, Appendix A3
I	Electrical current, A; Chapter 7
I	Electrical current semi-amplitude, A; Chapter 8
$I_{\min}$	Lower limit of linear region of power unit, A; Chapter 8
$\mathcal{I}$	Imaginary part; Equations (4.50) and (4.52); Chapters 4, 7
i	$= \sqrt{-1}$
i	Instantaneous electric current, A; Chapter 8
j	A number; Chapters 7, 9
$\mathbf{K}$	*Mass transfer:* equilibrium constant = $(v_j)_{\text{interface}}/(v_1)_{\text{interface}}$, dimensionless. *Heat transfer:* F_j/F_1, dimensionless; Chapters 4, 7, 9, Appendix A8

K	Rate constant, dimensionless, Chapter 5
K	Defined by Equations (9.17) and (9.18), dimensionless; Chapter 9
K_0	Modified Bessel function of the second kind; Appendix A3
k	Proportionality constant in Figure 8.12; Chapter 8
k	A constant, Appendix A4
k	First-order chemical reaction constant, sec^{-1}; Chapter 5
$\mathbf{k}$	Thermal conductivity, cal sec^{-1} °C^{-1} cm^{-1}; Chapters 6, 7
k, k'	Transfer coefficients for mass, or for heat ($=h/F_1$), cm sec^{-1}. Defined by: $\tilde{q} = k_1 v_1 - k_j v_j/\mathbf{K}$ $= k_1'\mathbf{K}v_1 - k_j'v_j$; Chapters 4, 7, 9, Appendix A8
L	Inductance per unit length, H cm^{-1}; Chapters 1, 7
L	Length of wire in a heater, cm; Chapter 8
L	Length of bed or physical system, cm
L	Reference length, cm; Chapters 4, 5
L	Limit of integral; Appendix A3
l	Length, cm; Chapters 7, 8
l	Depth of pocket (dead space), cm; Chapter 9
l_j	Length of a channel making up a porous medium, cm; Chapter 9
l_n	Length of nth path, cm; Appendix A10
l_1, l_2	Bed lengths when obtaining response by difference, cm; Chapter 6
M	Number of parameters in model, dimensionless; Chapter 9
M	Equation (5.8); Chapter 5

M	Thermal capacity, cal deg^{-1}; Chapter 8
M	Total flow rate, $cm^3\ sec^{-1}$; Appendix A10
$\tilde{M}$	Equation (5.8); Chapter 5
m	$= r\omega'$, dimensionless; Chapter 7
m	Ratio of resistances in thermometer bridge, dimensionless; Chapter 8
m	Number of heaters in parallel, dimensionless; Chapter 8
m	Period divisor, dimensionless; Chapter 8
m_n	Flow rate in nth path, $cm^3\ sec^{-1}$; Appendix A10
m_n	nth time moment about the mean m_1, mal $cm^{-3}\ sec^{n+1}$; Chapters 4, 5, Appendix A7
m_n^*	nth time moment about any origin, mal $cm^{-3}\ sec^{n+1}$; Chapter 4, Appendix A7
m_0	Zeroth moment, Chapters 4, 9, Appendix A7
$\tilde{m}_1$	Difference between values of m_1/m_0 at two stations, Chapter 4
N	Number of parameters to be simultaneously found, dimensionless; Chapter 9
N	Number, Appendix A7
N	Biot number, $lk/a_j\mathbf{K}$, dimensionless; Chapter 7
N_f	Frequency number, $L\omega/U$; Chapter 9, Appendix A10
n	A number ($=r_j$) designating number of parameters; Chapter 9
n	$= r_0\omega'$, dimensionless; Chapter 7
n	Ratio of resistances in thermometer bridge, dimensionless; Chapter 8
n	Period multiplier, dimensionless; Chapter 8
n_{ji}	Number of channels making up a porous medium of length l_j and radius r_i, dimensionless; Chapter 9
n_p	Fraction of v_0; $p = a, b$; dimensionless; Chapter 3
$\tilde{P}$	$= (1 + \mathrm{Pe}^2)^{1/2}$; Chapter 3
Pe	Peclet number UL/D, dimensionless
$\dot{p}_m$	$= [\frac{1}{4} + (K/\mathrm{Pe}_m)]^{1/2}$; Chapter 5
p_r	$= [\frac{1}{4} + (\tilde{s}/\mathrm{Pe}_r)]^{1/2}$; Chapter 5
p_v	Percentage error due to truncation; Chapter 9
$p(s)$	Equation (4.13), dimensionless; Chapter 4
p_1	Percentage error proportional to instantaneous value of concentration; Chapter 9
p_2	Percentage error (fixed amount); Chapter 9
Q	Longitudinal flux phasor, resultant at any point, mal $sec^{-1}\ cm^{-2}$; Chapters 6, 7
Q	Instantaneous source; see Table 2.1; Chapters 2, 3
Q	Flow rate out of sine-wave generator, $cm^3\ sec^{-1}$; Chapter 8
Q^+	Longitudinal flux phasor, incident wave at exit boundary, mal $sec^{-1}\ cm^{-2}$; Chapter 6
Q^-	Longitudinal flux phasor, reflected wave at exit boundary, mal $sec^{-1}\ cm^{-2}$; Chapter 6
Q_{in}	Flow rate into sine-wave generator, $cm^3\ sec^{-1}$; Chapter 8
Q_j	Flux, phase vector, per unit volume of flowing fluid from jth phase to flowing phase, mal $sec^{-1}\ cm^{-3}$; Chapters 1, 7
Q_{R}	Longitudinal flux phasor, resultant at exit boundary, mal $sec^{-1}\ cm^{-2}$; Chapter 6
Q_{s}	Longitudinal flux phasor, resultant at entrance boundary, mal $sec^{-1}\ cm^{-2}$; Chapter 6

q	Longitudinal flux, instantaneous value, mal sec^{-1} cm^{-2}; Chapters 4–7
q	Flow rate, instantaneous value, cm^3 sec^{-1}; Chapter 8
q	Continuous source; see Table 2.1; Chapter 2
q_j	Instantaneous flux (per unit volume of phase 1) from jth reservoir phase, mal sec^{-1} cm^{-3}; Chapters 1, 6, 7, 9, Appendix A8
$\tilde{q}_j$	Instantaneous flux per unit area of interface between reservoir phase j and flowing phase, mal sec^{-1} cm^{-2}; Chapters 1, 6, 7, 9, Appendix A8
q^+	Longitudinal flux, instantaneous value, incident wave, mal sec^{-1} cm^{-2}; Chapter 6
q^-	Longitudinal flux, instantaneous value, reflected wave, mal sec^{-1} cm^{-2}; Chapter 6
$\mathbf{q}^+$	Longitudinal flux phasor, incident wave, mal sec^{-1} cm^{-2}; Chapter 6
$\mathbf{q}^-$	Longitudinal flux phasor, reflected wave, mal sec^{-1} cm^{-2}; Chapter 6
R	$= x^2 + y^2 + z^2$; Chapter 2
R	Total resistance of potentiometer, ohm; Chapter 8
R	Resistance of electrical heater, ohm; Chapter 8
R	Resistance (electrical) per unit length, ohm cm^{-1}; Chapters 1, 7
R	Resistance in any arm of bridge, ohm; Chapter 8
R	$= \tilde{V}/k\tilde{a}$, resistance to mass or heat transfer, sec; Chapters 1, 7, Appendix A8
R_L	Resistance of load, ohm; Chapter 8
$\mathcal{R}$	Real part, Equations (4.50), (4.52), (5.7), (5.8); Chapters 4, 5
r	$= x^2 + y^2$; Chapter 2
r	Resistance of part of rheostat or potentiometer, ohm; Chapter 8
r	Number of unknown parameters defining Y_1 and Y_2, dimensionless; Chapter 9
r_g	Resistance of detector in bridge, ohm; Chapter 8
r_0	Radius of sphere, cm; Chapters 6, 7
r_0	Radius of cylindrical conduit, cm; Appendix A10
r_1	Reference resistance of thermometer, ohm; Chapter 8
$S(\omega)$	Complex or vector frequency content of a pulse; Chapter 4
$\mathrm{Si}(N_f)$	Sine integral of N_f; Appendix A10
s	Specific electrical resistance, ohm-cm; Chapter 8
s	Laplace variable
s	Defined by Equation (6.25), cm^{-1}; Chapter 6, Appendix A9
s^*	As s, for j stationary sources of flux; Equation (9.4); Chapter 9
$\tilde{s}$	$= st_0$, dimensionless Laplace variable; Chapter 5
T	$= (1 + W^2)^{1/2}$, dimensionless; Chapter 6
T	Multiplying factor; Chapter 9
T	Temperature, deg; Chapter 8, Appendix A8
T_n	Temperature uncertainty, deg, Chapter 8
T_w	Duration of pulse (Figure 4.4), sec; Chapter 4
t	Time, sec
t	Limit of integral, Appendix A3
t_0	Origin of time; Chapter 1
t_1	$= t_0 + (\frac{1}{4})\,\tilde{t}$; Chapter 1

t'	Time whose origin is $-\infty$ on scale of t, sec; Chapter 3
$\bar{t}$	Mean time of a distribution, sec; Chapters 3, 5, 9
$\tilde{t}$	Period of a sine wave, sec; Chapters 1, 6, 8
t_{lag}	Difference in phase of two sine waves, sec; Chapters 1, 8
t_{peak}	Time at peak of a pulse, sec; Chapters 3, 9
t_0	$= L/U$, sec; Chapters 4, 5
$t_{1/2}$	Time that halves area under pulse, sec; Chapters 3, 9
U	Velocity of fluid, actual, cm sec^{-1}
U_{max}	Maximum velocity in steady Poiseuille flow, cm sec^{-1}; Appendix A10
V	Voltage applied to bridge, V; Chapter 8
V	Voltage applied to heater, V; Chapter 8
V	Voltage applied to circuit, V; Chapter 8
V	Concentration phasor, resultant at any point, mal cm^{-3}, Chapters 1, 6, 7, 9
$\tilde{V}$	Semiamplitude displaced volume of sine-wave generator reservoir, cm^3; Chapter 8
V^+	Concentration phasor, incident wave at exit boundary, mal cm^{-3}; Chapter 6, Appendix A10
V^-	Concentration phasor, reflected wave at exit boundary, mal cm^{-3}; Chapter 6
$\tilde{V}_j$	Volume of jth phase per unit volume of flowing phase, dimensionless; Chapters 6, 7, 9, Appendix A8
V_{ji}	Amplitude of sine wave in the jith channel, mal cm^{-3}; Chapter 9
V_{L}	Voltage across load, V; Chapter 8
$\tilde{V}_{ji}$	Volume of n_{ji} channels making up a porous medium, of length l_j and radius r_i, cm^3; Chapter 9, Appendix A10
$\tilde{V}_{\text{m}}$	Volume of mixer, cm^3; Chapter 8
V_{R}	Concentration phasor, resultant at exit boundary, mal cm^{-3}; Chapter 6
V_{R}	Voltage across potentiometer, V; Chapter 8
V_{S}	Concentration phasor, resultant at entrance boundary, mal cm^{-3}; Chapter 6
V_{w}	Velocity of wave or pulse, cm sec^{-1}; Chapters 1, 4
$\bar{V}_j(t)$	Phase vector of mean concentration in jth phase, mal cm^{-3}; Chapters 6, 7, 9
$\tilde{V}_0$	Volume of sine-wave generator reservoir; standing value, cm^3; Chapter 8
$\mathbf{V}$	Amplitude of net sine wave issuing from bundle of channels, mal cm^{-3}; Chapter 9, Appendix A10
v	Concentration of heat ($=F \times$ temperature) or matter, instantaneous value, mal cm^{-3}
v	Programming voltage, instantaneous value, V; Chapter 8
$\tilde{v}$	Volume of sine-wave generator reservoir, instantaneous value, cm^3; Chapter 8
v_a	Voltage, reference value, V; Chapter 8
v_e	Voltage, error signal, V; Chapter 8
$v_{\text{interface}}$	Concentration at interface surface of reservoir, mal cm^{-3}; Chapter 4
v_m	Voltage, feedback, V; Chapter 8

v_{trunc}	Value of concentration at point of truncation of pulse, mal cm^{-3}; Chapter 9	x	Spatial dimension, cm; Chapter 2
v_{ex}	Simulated "experimental" value of concentration, mal cm^{-3}; Chapter 9	x	$= a[\lambda + (b/a\lambda)]$; Appendix A3
v_0	Asymptotic value of concentration in degenerate step change, mal cm^{-3}; Chapter 3	$Y(s)$	$= \{\ln [F(s)]^{-1}\}^{-1}$; Chapter 4
v^+	Concentration, instantaneous value, incident wave, mal cm^{-3}; Chapters 6, 9, Appendix A10	$\mathbf{Y}$	Complex shunt admittance $(= Y_1 + iY_2)$ of reservoir phase, sec^{-1} (in thermal or concentration wave systems); Chapters 6, 7, 9, Appendix A8
v^-	Concentration, instantaneous value, reflected wave, mal cm^{-3}; Chapter 6	y	General variable; Chapter 1, Appendix A4
$\bar{v}(s)$	Laplace transform of concentration, mal cm^{-3}; Chapter 9	y	Spatial dimension, cm; Chapter 2
$\bar{v}_j(t)$	Spatial average concentration of jth reservoir phase, mal cm^{-3}; Chapters 7, 9	y_0	Reference value; Appendix A4
$\mathbf{v}^+$	Concentration phasor, incident wave, mal cm^{-3}; Chapters 1, 6, 9, Appendix A10	Z	$= z - Ut$, cm; Chapters 2, 3, Appendix A2
$\mathbf{v}^-$	Concentration phasor, reflected wave, mal cm^{-3}; Chapters 1, 6	Z	Spatial variable, cm; Chapter 2
W	Watts dissipated by resistance thermometer, W; Chapter 8	Z	Impedance (= 1/shunt admittance) of reservoir phase, sec; Chapter 7, Appendix A8
W	$= 4\omega D/U^2$, dimensionless; Chapter 6	Z	Impedance to longitudinal wave, sec cm^{-1}; Chapters 1, 6
w	General variable; Appendixes A1, A2	Z^+	Impedance to incident longitudinal wave, sec cm^{-1}; Chapter 6
X_n	A quantity; Appendix A7	Z^-	Impedance to reflected longitudinal wave, sec cm^{-1}; Chapter 6
$\bar{X}$	Mean value; Appendix A7	z	A general variable; Appendix A4
$X(s)$	$= s\{\ln [F(s)]^{-1}\}^{-2}$; Chapter 4	z	Distance in direction of flow, cm
x	General variable; Chapter 1, Appendixes A1, A2	z'	$= -z$; distance against direction of flow, cm; Chapter 1

Greek Letters

α	Temperature coefficient of resistivity, deg^{-1}; Chapter 8
α	$= D_1/D_2$, ratio of dispersion coefficients in regions 1 (fore section) and 2 (central section), dimensionless; Chapter 5
α	A variable; Appendix A9
α	Attenuation constant, cm^{-1}; Chapters 1, 6
α_1	Attenuation constant for incident longitudinal wave, cm^{-1}; Chapters 1, 6
α_2	Attenuation constant for reflected longitudinal wave, cm^{-1}; Chapters 1, 6
β	Phase constant for incident and reflected longitudinal waves, cm^{-1}; Chapters 1, 6
β	$= D_3/D_2$, ratio of dispersion coefficients in regions 2 (central section) and 3 (tail section), dimensionless; Chapter 5
Γ	Defined by Equation (7.39), dimensionless; Chapter 7
Γ	Gamma function; Appendix A4
Γ	Reflection coefficient, dimensionless; Chapter 6
γ	Propagation constant for longitudinal waves, cm^{-1}; Chapters 1, 6
γ_1	Propagation constant for incident longitudinal wave ($= \alpha_1 - i\beta$), cm^{-1}; Chapters 1, 6
γ_2	Propagation constant for reflected longitudinal wave ($= \alpha_2 + i\beta$), cm^{-1}; Chapters 1, 6
Δ	Difference or interval; Chapters 3, 4, 7
$\delta, \delta(t), \delta(z)$	Dirac delta function; Chapters 5, 8
ε	Defined by Equation (5.50); Chapter 5
ε	Porosity (fractional voidage), dimensionless
ε_{int}	Fraction of volume of sphere occupied by pores, dimensionless; Chapter 7
ε_{ext}	Fraction of packed bed occupied by flowing medium, dimensionless; Chapter 7
ζ	Defined by Equation (7.35), dimensionless; Chapter 7
ζ	$= \chi - \chi_m$, dimensionless distance; Chapter 5
ζ	$= \xi/(2D^{1/2}t^{1/2})$; Appendix A3
ζ	$= k\xi$; Appendix A4
η_{ji}	Analytic function of l_j and r_i; Equation (9.26); Chapter 9
Θ	$= (1 - \theta)/\theta^{1/2}$, dimensionless; Chapter 3
θ	$= tU/L$, dimensionless time; Chapters 3, 5
θ	$= t - (z/U)$; Appendix A1
κ_1	$= \tilde{a}_j k_1/\tilde{V}_j$, sec^{-1}; Chapters 4, 7, Appendix A8
κ_j	$= \tilde{a}_j k_j/\tilde{V}_j\mathbf{K}$, sec^{-1}; Chapters 4, 7, Appendix A8
κ_1'	$= \tilde{a}_j k_1'\mathbf{K}/\tilde{V}_j$, sec^{-1}; Chapters 4, 7, Appendix A8
κ_j'	$= \tilde{a}_j k_j'/\tilde{V}_j$, sec^{-1}; Chapters 4, 7, Appendix A8
Λ	Defined by Equation (7.30), dimensionless; Chapter 7
λ	$= b/a\xi$; Appendix A3

λ — Wavelength, cm; Chapters 1, 6, 8

μ — Equation (4.12), dimensionless; Chapter 4

$\mu(\omega)$ — $= [\Pi(\omega) - \nu(\omega)/\mathrm{Pe}]U/L$, Equation (9.22), $\sec^{-1}$; Chapter 9

μ_r — Equation (5.54), dimensionless; Chapter 5

$\dot{\mu}_m$ — Equation (5.65), dimensionless; Chapter 5

$\bar{\mu}_n$ — Laplace transform of normalized nth time moment about the mean; Equation (4.39); Chapter 4

ν — Dimensionless concentration; Chapters 3, 5

ν — $= \Pi^2 - \psi^2$, dimensionless; Chapter 9

ν_r — Equation (5.54); Chapter 5

$\dot{\nu}_m$ — Equation (5.66); Chapter 5

Ξ — Defined by Equation (7.31), dimensionless; Chapter 7

ξ — A variable [see Section (2.3.3)]

ξ — A variable; Chapter 3, Appendix A3

ξ — Representing t or Θ; Chapter 3

ξ — A variable; Appendix A4

Π — $= \ln[V^+(L)/V_S(0)]$; Chapter 9

π — Generalized parameter defining Y_1 and Y_2, dimensioned or dimensionless; Chapter 9

π^3 — Third normalized moment about the mean, $\sec^3$; Chapter 4

ρ — Density, g cm^{-3}; Chapters 5, 7, 9, Appendix A8

Σ — Defined by Equation (7.34), dimensionless; Chapter 7

σ — Defined by Equation (7.36), dimensionless; Chapter 7

σ^2 — Variance, $\sec^2$ or dimensionless; Chapters 3, 4, 5, 9, Appendixes A4, A7

ϕ — Phase angle at a point in a reservoir phase, or of the flux, relative to the wave in the flowing fluid, rad or deg; Chapter 7

ϕ — A function; Appendix A1

ϕ — $= \tan^{-1}(Y_2/Y_1)$; Chapter 9

ϕ_0 — Defined by Equation (7.42), dimensionless; Chapter 7

$\phi_1(r)$ — Defined by Equation (7.43), dimensionless; Chapter 7

ϕ_2 — Defined by Equation (7.44), dimensionless; Chapter 7

χ — $= z/L$, dimensionless distance; Chapter 5

χ — $= z/Ut$, dimensionless distance; Chapter 3

ψ — Phase angle or epoch at distance z, rad; Chapters 1, 4, 6, 8, 9, Appendix A10

ψ_{ji} — Phase angle of wave in the jith channel; Chapter 9, Appendix A10

ψ_r — Equation (5.57e); Chapter 5

ψ_r' — Equation (5.57f); Chapter 5

Ψ — Phase angle of net wave issuing from bundle of channels; Chapter 9, Appendix A10

ω — Radial frequency, rad $\sec^{-1}$

ω' — $= (\omega/2a)^{1/2}$, cm^{-1}; Chapters 1, 6, 7

Subscripts

a	Value at measuring station a; Chapter 4	peak	Time at highest point of pulse; Chapter 3
a	Value in region a; Chapter 6	pl	Plane (source); Chapters 2, 3
a	Defining value of y, z, I; Appendix A4	pt	Point (source); Chapter 2
b	Value at measuring station b; Chapter 4	q	Value at source; Chapter 2
b	Value in region b; Chapter 6	r	Designating rth region or boundary; Chapter 5
c	Value at station c; Chapter 4	r	Denoting distance on error-function curve; Chapter 3
c	Admittance based on concentrations; Appendix A8	T	Thermometer; Chapter 8
D	Characteristic length; Chapter 2	T	Admittance based on temperature; Appendix A8
d	Value at station d; Chapter 4	t	Variance in units of $(\text{time})^2$; Chapters 3, 5
i	Concentration at inlet; Chapter 4	x	Direction, Chapter 1
interface	Concentration at $j-1$ interface (in either phase); Appendix A8	y	Direction; Chapter 1
$j(=2, 3, \ldots)$	Reservoir phase	θ	Variance in dimensionless time; Chapters 3, 5
$j(=1)$	Flowing phase	0	Station, Chapter 5
L	Leads; Chapter 8	0	Value as $s \longrightarrow 0$; Chapter 4
l	Line (source); Chapter 2	0	Defining value of y; Appendix A4
m	Measuring station; Chapter 5	$\frac{1}{2}$	Time that divides pulse into two equal areas; Chapter 3
$m(=1, 2)$	A number; Chapter 6	1	Phase 1 (flowing); Chapters 6, 7, Appendix A8
n	nth moment; Chapter 4	1	Region or station; Chapters 2, 5
n	Defines integral; Appendix A3	1	Particular value of θ in normal error curve; Chapter 3
n	Path of interfering wave; Appendix A10	1	Identifying transport or equilibrium coefficient in a "series model"; Chapters 7, 9
o	Value at exit ("out"); Chapter 4	2	Identifying transport or equilibrium coefficient in a "series model"; Chapters 7, 9
p	Pressure; Appendix A8		

Superscripts

$'$, $''$	Differentiation once, twice with respect to s; Chapter 4	*	Moments about the mean
*	Desired (computed final) value of parameter; Chapter 9		

Barred quantities generally are transformed (Laplace, Fourier) variables or spatial averages. Double bars indicate transforms of spatial averages.

Chapter 1

Introduction and Waves and Pulses

Ranging from pressing a thumb on new paint to pushing forward the control column of an aircraft, the process of learning of the properties and behavior of anything by altering its rest state is widely used.

The subject has several aspects; first of all, the thing being examined may be an existing one—e.g., paint, aircraft, distillation column, the human body, or even a country's economy—or, on the other hand, it may be an artificial one, specially set up in the laboratory. Second, the disturbance from the rest state may be one of several kinds—it may be a sharp impulse, a steady oscillation, or it may be accidental, the kind of change that often occurs in a random way. Third, the study may have either of two purposes, viz., to find out how the system reacts to a change (whether and how it tends to keep on changing or to restore itself), or to measure the numerical value of quantities. In the latter case the disturbance is only a means to an end.

Finally, the mathematics can range from the use of analytic solutions of linear equations in order to produce unambiguous results to the use of statistical and optimization techniques on a lot of information obtained from a complicated system. Furthermore, although the disturbance is one that changes with time, the analysis does not have to involve time as a variable. In addition the change with *distance* may also be informative. The point is illustrated in Figure 1.1, where (a)–(c) represent any system (mechan-

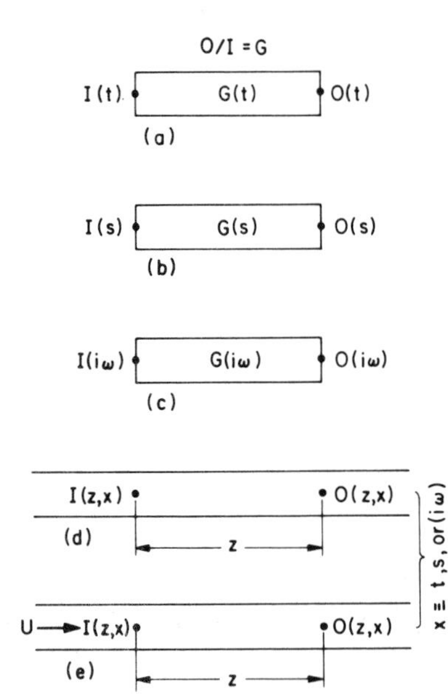

Fig. 1.1. The response of a bounded system *and* traveling waves in an extended medium.

ical, electrical or electronic, physiological, chemical, physical) which is disturbed by an input I, and the response of the system O is measured. These are labeled as though there actually were both an input and an output, in contrast to Figure 1.1(d, e). The ratio of the two is called the *response* $G = O/I$. These changes may be written as functions of time [Figure 1.1(a)], or they may be Laplace-transformed to be functions of the Laplace variable s [Figure 1.1(b)], or they may be transformed by the Fourier transform to be functions of sine waves of frequency ω [Figure 1.1(c)].

In contrast, Figure 1.1(d) shows a system in which there is no clearly defined input and output, but changes are occurring at all points in one (or two or three) spatial dimensions, and so the changes at any two positions a distance z apart may be thought of either as being an input and output, or as being the changes due to a wave or pulse which is moving in space in a system that is of physical size—limited or unlimited. It is the latter idea that is expanded in this book and the change in *something* (to be discussed next) would primarily be in distance and time. It will, however, be examined as a function of distance and either time or the transform variable s or the frequency ω. In other words, the treatment will deal with a traveling wave or pulse, almost wholly in one spatial dimension. The difference between this treatment and that of other traveling wave systems is that the medium itself is moving at a steady velocity relative to points I and O, as shown in Figure 1.1(e). This seemingly small change has large consequences, while it permits many quantities to be measured.

The Wave

The thing that changes in the wave will be either the concentration of material (dissolved in liquid or mixed in gas) or the temperature of the fluid. The basic equations are very similar for both; to avoid having to describe each separately, it is proposed to use concentration as the thing that changes, i.e., the dependent variable.

A symbol v has been chosen (although Carslaw and Jaeger [2] use v to indicate temperature). Thus, either: $v \equiv$ concentration of matter, in mass (or mol) per unit volume (e.g., mol cm^{-3}), or $v \equiv$ concentration of heat (e.g., cal cm^{-3}), e.g, $v = F \times$ temperature, where F is the volumetric heat capacity (e.g., cal cm^{-3} deg^{-1}). This leads to great convenience in the mathematics, but some uncertainty may arise when temperature is involved. (If in doubt, one should write the equations in terms of temperature.) This uncertainty may exist when F changes.

The situation at the interface is assumed, as is usual, to be governed by equilibrium thermodynamics, and the similarities between mass and thermal systems are shown in Figure 1.2. An interface is shown, either between two phases, where the flow is assumed parallel to the interface, or in the one phase, where the flow is through the interface. The thermodynamic equilibrium conditions are: (i) in mass transport systems the fugacity has no discontinuity across the interface; (ii) in thermal systems the temperature has no discontinuity. In both systems, however, the *concentration* (of matter or heat) may show a discontinuity, and the ratio of these concentrations will be denoted by $\mathbf{K}$.

Later in this chapter some wave equations will be discussed, while their treatment for various situations and purposes forms the rest of the book. When the actual physical situation is considered it might be surprising that such relatively simple equations have proved to be as satisfactory as they have, for each flowing parcel of fluid is usually subjected to random motion in time and space, in irregular channels. [Thus, although at the inlet of a

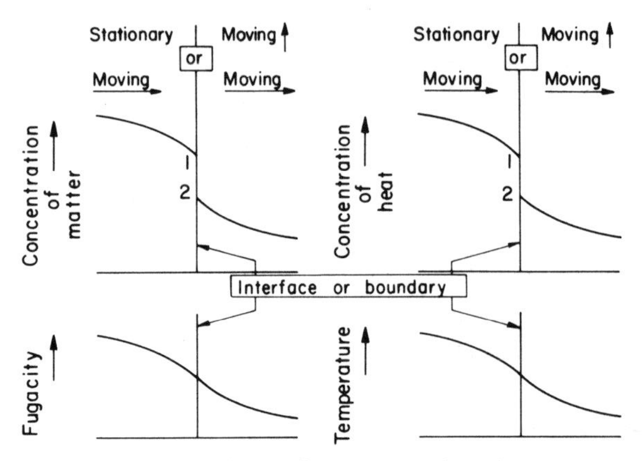

Fig. 1.2. Concentrations across a boundary.

system (whatever that is) a disturbance is considered to be made up of *coherent* waves, at the outlet where the wave is measured, because of the different histories of each part of the fluid, the waves arrive as *interfering* but still *coherent* waves; see general texts on waves and Appendix 10.] The measured effect is thus an average of all these contributing causes. However, the numerical value of the quantity determined in the laboratory, e.g., heat transfer coefficient, or mean velocity of flow, may well be used in design problems if the physical experiment simulates the actual system and if changes are such that linearity is a fair assumption.

1.1 WAVES AND PULSES: GENERAL

A wave in any medium is the transmission of some disturbance of the normal rest state of the medium from one place to another; the medium suffers no permanent effect but there is a flow of some kind associated with the wave—of energy, heat, or "chemical" matter, i.e., material dissolved in a medium that is not moved by the wave.

Waves have been studied for well over 200 years, and the investigation—especially of the behavior of electromagnetic waves—is now more intense than ever. Much more recent, however, is the study of waves or pulses in a medium that is itself moving relative to the generator of the wave and to the observer—and often, also, relative to something that affects the wave from outside the flowing medium. For convenience (and usually, in this book) a *wave* may refer to a *repetitive* fluctuation, a *pulse* to a *single* fluctuation, and an *impulse* to the mathematical, limiting, concept of a *Dirac δ-function*, which is impossible to generate but makes for some ease of analysis.

These fluctuations will be in nonmechanical quantities—temperature or the concentration of dissolved material—and to emphasize this distinction, Lighthill and Whitham [1] called them *kinematic*. A study could be directed either toward calculating the effect of a specified system on a wave or pulse, or toward deducing the values of parameters, or toward elucidating the system that produces a certain (measured) effect. The second and third inspire the theme of this book. The consequences are that the system (i.e., the *mathematical model*) can be as simple as is thought to be consistent with the facts, the wave or pulse can be chosen to give the best balance between experimental and mathematical facility, but, in contrast, the penalty is that the experiment has to be set up to simulate the mathematical model, as well as to fulfill any conditions (such as the accuracy of measurement) imposed by the analysis.

In general, the disturbances to be considered in this book are those that allow of fairly precise description, viz., the Dirac impulse, a humplike pulse describable with some accuracy by its lower moments and (particularly) a steady sinusoidal wave. (Other pulses or waves may be resolved by harmonic analysis.)

The systems being considered will allow some of the following quantities to be measured: heat transfer coefficient, mass transfer coefficient, diffusion coefficient, diffusivity, conductivity, surface diffusivity, equilibrium partition coefficient, longitudinal dispersion (i.e., mixing) coefficient, the velocity of flow of the medium, and the distribution of sizes of certain structures. In addition, the effectiveness of a chromatographic (i.e., adsorption) process, and the quantities associated with simple flow structures may be deduced. The subject of residence-time distributions is not specifically covered here, being catered for in other works, e.g., Levenspiel [2].

1.2 ONE-DIMENSIONAL TRAVELING WAVES

If some quantity v depends upon both distance z and time t such that

$$v = f(V_w t - z) \tag{1.1}$$

then if $v = v_a$ at $z = z_a$ and $t = t_a$, v can be v_a again when, simultaneously, $z = z_a + \lambda$ and $t = t_a + \tilde{t}$ if

$$\lambda - V_w \tilde{t} = 0 \tag{1.2}$$

as can easily be seen by increasing z and t appropriately in Equation (1.1), which can thus be considered to be a description of a traveling wave as in Figure 1.3. The wave can be considered to be traveling with a *velocity* V_w that satisfies Equation (1.2). If the wave in Figure 1.3 were repeated as

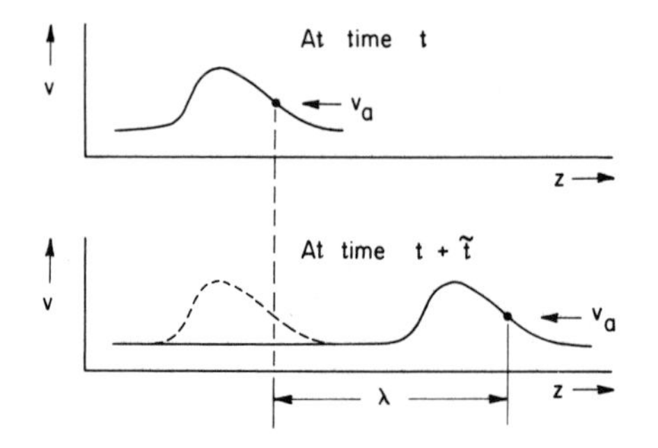

Fig. 1.3. Traveling pulse and repetitive waves.

shown dotted so that the new wave is in the position of the old at time $\tilde{t}$ later, when the first had advanced through a distance λ, then λ is called the *wavelength* and $\tilde{t}$ is called the period of the (periodic) wave. Further, from Equation (1.2)

$$V_w = \lambda/\tilde{t} \tag{1.3}$$

If the function were of the form

$$v = F(V_w t + z) \tag{1.4}$$

then the same argument applies except that this would represent a wave traveling in the opposite direction to that in Equation (1.1).

Equations (1.1) and (1.4) are roots of a differential equation, obtained from an appropriate balance over an infinitesimal distance dz—a balance of forces in the case of sound waves, for example, or of heat or material in the case of thermal or diffusional waves.

The heat or mass balance for a wave in a flowing fluid is derived by taking a heat or mass balance over this differential length. However, it is sometimes convenient, in all wave systems, to use the (analogies of the) telegrapher's equations. Of these there are two for each system; both relating a "flow" to a "driving force." Many examples will be found in Moore [3], while Chapter 6 derives and discusses them for the system being considered in this book. In every case, elimination of either the "force" or the "flow" variable between the two telegrapher's equations gives rise to the differential equation in the remaining variable.

1.3 THE WAVE EQUATION

The above-mentioned differential equation is called the *wave equation*; Coulson [4] suggests it be called the *equation of wave motion* to avoid confusion with the term *wave equation* of wave mechanics. (However, the word "motion" introduces its own difficulties when kinematic waves are being dealt with.)

It is first or second order in the *space* coordinates in this chapter but it may be of higher order; see Chapter 7. It may be first or second order in *time* also; the consequences of the differences in either variable are radically different. In general, only one-dimensional flow, and mainly one-dimensional temperature or concentration distributions, are considered in this book. The wave equations that arise in this system can be compared with three others, and this is done in Table 1.1, which thus contains seven equations describing different systems. These are commented on now.

TABLE 1.1

WAVE EQUATIONS FOR IMPORTANT FLOW AND NONFLOW SYSTEMS

Flowing medium

$$\frac{\partial^2 v}{\partial z^2} - 0 - \frac{1}{D}\left(\frac{\partial v}{\partial t} - q_j\right) - 0 - \frac{U}{D}\frac{\partial v}{\partial z} = 0 \qquad (1.5a)$$

$$\frac{\partial^2 v}{\partial z^2} - 0 - \frac{1}{D}\left(\frac{\partial v}{\partial t} - 0\right) - 0 - \frac{U}{D}\frac{\partial v}{\partial z} = 0 \qquad (1.5b)$$

$$0 - 0 - \left(\frac{\partial v}{\partial t} - q_j\right) - 0 - U\frac{\partial v}{\partial z} = 0 \qquad (1.5c)$$

$$0 - 0 - \left(\frac{\partial v}{\partial t} - 0\right) - 0 - U\frac{\partial v}{\partial z} = 0 \qquad (1.5d)$$

Stationary medium

$$\frac{\partial^2 v}{\partial z^2} - 0 - \frac{1}{a}\frac{\partial v}{\partial t} - 0 - 0 = 0 \qquad (1.5e)$$

$$\frac{\partial^2 v}{\partial z^2} - 0 - 0 - \frac{1}{c^2}\frac{\partial^2 v}{\partial t^2} - 0 = 0 \qquad (1.5f)$$

$$\frac{\partial^2 v}{\partial z^2} - RGv - (RC + LG)\frac{\partial v}{\partial t} - LC\frac{\partial^2 v}{\partial t^2} - 0 = 0 \qquad (1.5g)$$

Flow systems [Equations (1.5a)–(1.5d); $v \equiv$ temperature or concentration of heat ($F \times$ temperature) or matter].

Equation (1.5a) has finite dispersion (i.e., mixing in the flow stream), together with a supply q_j of heat or matter from a reservoir phase outside the flowing medium.

Hence, $U \neq 0,\ D \neq 0,\ q_j \neq 0$.

Equation (1.5b) is the same, but $q_j = 0$.

Hence, $U \neq 0,\ D \neq 0,\ q_j = 0$.

Equation (1.5c) has no dispersion in the flowing medium, but has an external reservoir phase.

Hence, $U \neq 0,\ D = 0,\ q_j \neq 0$.

Equation (1.5d) has neither dispersion in the flowing medium nor an external supply.

Hence, $U \neq 0,\ D = 0,\ q_j = 0$.

Nonflow systems [Equations (1.5e)–(1.5g)].

Equation (1.5e) is for a *dissipative system* (such as thermal conduction or mass diffusion systems) in which the wave suffers severe attenuation as it proceeds. In the first case v may be temperature or concentration of heat, with a the thermal diffusivity, while in the second case v is concentration and a is the diffusion coefficient.

Hence, $U = 0,\ a \neq 0,\ q_j = 0$.

Equation (1.5f) is for a lossless system in which the wave suffers no attenuation of amplitude as it proceeds; c is a constant; it turns out to be the wave velocity and to be related to the parameters of the system. See Moore [3] for examples.

Hence, $U = 0,\ a = \infty,\ q_j = 0$.

[Equations (1.5e) and (1.5f) may be combined when the system is such that there is some loss in a fundamentally lossless process.]

Equation (1.5g) is that for a distributed-parameter electrical transmission line as shown in Figure 1.4. It contains series inductance L, series resistance

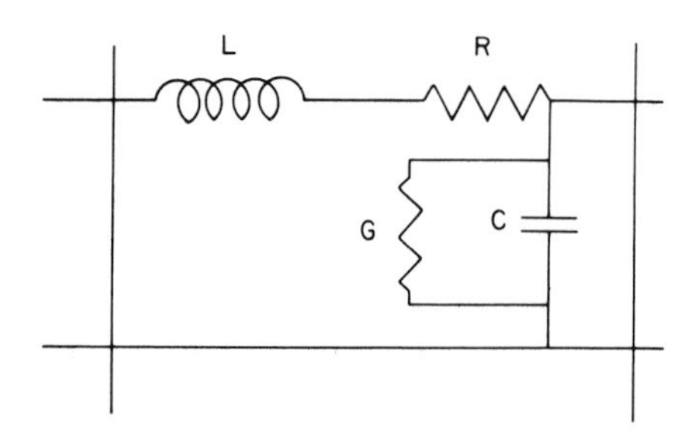

Fig. 1.4. Distributed-parameter electrical transmission line.

R, shunt conductance (i.e., reciprocal resistance) G, and shunt capacitance C, all amounts per unit length. Analogies will be drawn later between this last system and the flow system.

Sine Waves

The consequences of the various combinations of terms contained in Equations (1.5a)–(1.5g) become evident when solutions are obtained for certain situations, especially for the case of a steady, periodic, pure sine wave of radial frequency ω. The solution may be obtained either by finding

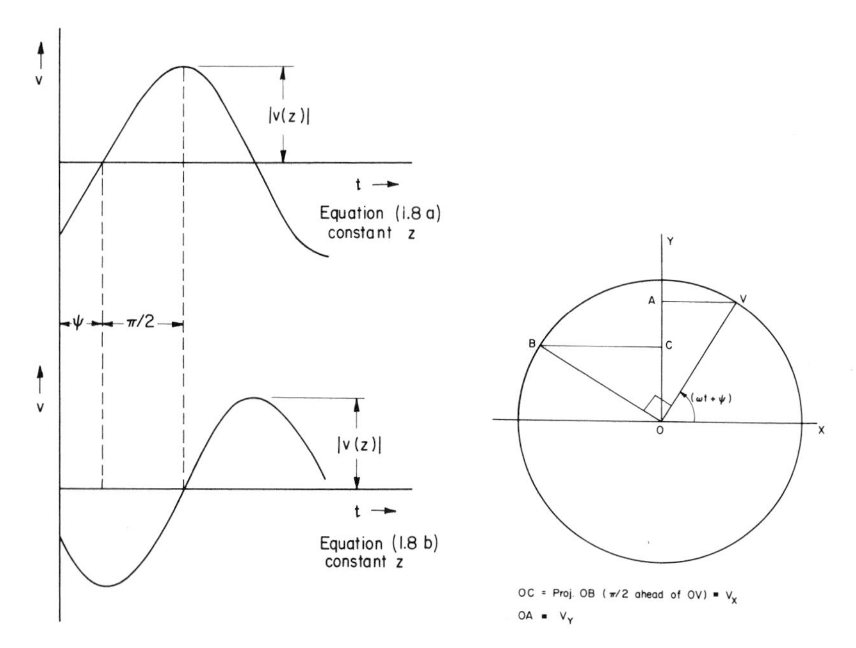

Fig. 1.5. The forms of Equations (1.8a) and (1.8b).

Fig. 1.6. Phasor (i.e., vector and phase angle) and components.

the Laplace transform and then substituting $i\omega$ for s, the transform variable, or t may be removed by substituting

$$v(z) = V(z)e^{i\omega t} \tag{1.6}$$

where $V(z)$ is a complex amplitude or phase vector of the sine wave; i.e., it will denote both half the swing of values of v at any value of z, and the *phase angle* of this wave relative to the wave at any other value of z. The use of $e^{i\omega t} = \cos \omega t + i \sin \omega t$ is for convenience; manipulation of the exponential form is easier, and in the solution one takes either the real or the imaginary part of the exponential to find the behavior of either a cosine wave *or* a sine wave. This procedure must not be followed blindly: in Chapter 7, for example, both real and imaginary terms must be retained in some of the equations; e.g., $\Re[V(x + iy)e^{i\omega t}]$ means $V' \cos \omega t$, where V' is a new complex semiamplitude of magnitude and phase angle, different from V, and is obtained by operating on V by $x + iy$. That is, $|V'|/|V| = (x^2 + y^2)^{1/2}$ and V' leads V (in the direction of ωt increasing) by $\tan^{-1}(y/x)$.

Again, neither concentration nor temperature can be negative in absolute values, but if values relative to some datum are taken (just as voltage is taken relative to the arbitrary datum of ground potential), then negative values may appear without conflicting with the physical realities.

Thus, from Equation (1.6)

$$v = |V(z)|\, e^{i\psi(z)} e^{i\omega t} = |V(z)|\, e^{i[\omega t + \psi(z)]} \tag{1.7}$$

meaning either

$$v_x = |V(z)| \cos(\omega t + \psi) = |V| \sin(\omega t + \psi + \pi/2) \tag{1.8a}$$

or

$$v_y = |V(z)| \sin(\omega t + \psi) \tag{1.8b}$$

which are functions of the form shown in Figure 1.5.

(i) *Geometric Interpretation*

The complex amplitude V can be resolved into its components, which are sometimes wanted, for

$$V = |V|\, e^{i\psi} = |V| (\cos\psi + i \sin\psi) = V_x + iY_y$$

It is thus specified completely by a vector of length $|V|$, making an angle ψ with a reference direction OX in Figure 1.6. The vector OV (or radius vector) is termed the *phase vector* or *phasor* in traveling wave engineering.

If the vector OV rotates at f revolutions per unit time then projections V_x and V_y vary sinusoidally and if angle $\widehat{VOX} = \psi$ at $t = 0$, then

$$\text{Projection} \quad V_x \equiv |V| \cos(\omega t + \psi) = v_x = |V| \sin(\omega t + \psi + \pi/2) = |V| \sin \widehat{XOB} \tag{1.9a}$$

and

$$\text{Projection} \quad V_y \equiv |V| \sin(\omega t + \psi) = v_y \tag{1.9b}$$

and the variation of temperature, concentration, or other appropriate variable is the same as the variation of length of the projections OC or OA of the revolving phasor OV.

Further, the difference between the two waves

$$\text{(a)} \qquad |V| \cos(\omega t + \psi) \equiv \Re[|V|\, e^{i(\omega t + \psi)}]$$

and

$$\text{(b)} \qquad |V| \cos \omega t \equiv \Re[|V|\, e^{i\omega t}]$$

is that the rotating phase vector of (a) is at an angle of ψ from the phase vector of (b), *measured in the same direction as has been arbitrarily chosen to be the direction of rotation of* $|V|$ *as* t *increases.*

Thus,

phasor (a) is ψ radians *leading* (b)

or

phasor (b) is ψ radians *lagging* (a)

Again, if the wave were written as $|V|e^{-i\omega t}$, then $|V|e^{-i(\omega t+\psi)}$ would be a sinusoidal variation leading it.

(*ii*) *Solution of Wave Equation: General Sine Wave*

If now it is specified that in a system governed by a wave equation such as Equations (1.5a)–(1.5g) the variable v at any value of z executes a sinusoidal variation with time, e.g., $v(z, t) = V(z)\cos\omega t$, then differentiation with respect to t and z and substitution into the appropriate equation in (1.5a)–(1.5g) will remove the time variable, giving a second-order ordinary differential equation in z. This is illustrated below; the differences in the auxiliary equations should be noted inasmuch as these now govern the behavior of $V(z)$ with distance.

Thus, from Equation (1.6)

$$\partial v/\partial z = e^{i\omega t}V', \qquad \partial^2 v/\partial z^2 = e^{i\omega t}V''$$
$$\partial v/\partial t = i\omega e^{i\omega t}V, \qquad \partial^2 v/\partial t^2 = -\omega^2 e^{i\omega t}V$$

where the prime refers to differentiation with respect to z and so Equations (1.5*a*)–(1.5g) become, respectively,

$$e^{i\omega t}[V'' - (U/D)V' - (1/D)(i\omega V - Q_j)] = 0 \tag{1.5a$'$}$$
$$e^{i\omega t}[V'' - (U/D)V' - (1/D)(i\omega V)] = 0 \tag{1.5b$'$}$$
$$e^{i\omega t}[UV' + (i\omega V - Q_j)] = 0 \tag{1.5c$'$}$$
$$e^{i\omega t}[UV' + i\omega V] = 0 \tag{1.5d$'$}$$
$$e^{i\omega t}[V'' - (i\omega/a)V] = 0 \tag{1.5e$'$}$$
$$e^{i\omega t}[V'' + (\omega^2/c^2)V] = 0 \tag{1.5f$'$}$$
$$e^{i\omega t}[V'' - RGV - i\omega(RC + LG)V + \omega^2 LCV] = 0 \tag{1.5g$'$}$$

where $q_j = Q_j e^{i\omega t}$. So, for all the second-order equations obtained by equating the square bracket to zero in Equations (1.5a′), (1.5b′), (1.5e′), (1.5f′), and (1.5g′), the solution giving the complex amplitude at z will be

$$V = Ae^{\gamma_1 z} + Be^{\gamma_2 z} \tag{1.10a}$$

where $\gamma_{\frac{1}{2}}$ are complex and are called the *propagation constants*; A and B are arbitrary constants, to be determined from other conditions of the problem.

[On the other hand, Equations (1.5c′) and (1.5d′) being first order, the solution will be

$$V = Ae^{\gamma z} \tag{1.10b}$$

i.e., $B = 0$ and $\gamma \equiv \gamma_1$. The value of γ_2 is discussed in Chapter 6.]

To return to $\gamma_{\frac{1}{2}}$, it is interesting to examine the differences between the above typical and important cases. [Equation (1.5a′) is not discussed here; it requires a relation between Q_j and V, which is the subject of Chapter 6.] Thus, we have the following:

For a flowing medium, from Equation (1.5b′),

$$\gamma_{\frac{1}{2}} = (U/2D) \pm [U^2/4D^2) + (i\omega/D)]^{1/2} \tag{1.5b''}$$

[Equations (6.30)–(6.37) continue the derivation.]

For a nonflow, lossy system, from Equation (1.5e′),

$$\gamma_{\frac{1}{2}} = \pm(\omega/2a)^{1/2}(1 + i) \tag{1.5e''}$$

For a nonflow, lossless system, from Equation (1.5f′),

$$\gamma_{\frac{1}{2}} = \pm i\omega/c \tag{1.5f''}$$

For a transmission line, from Equation (1.5g′) (the behavior depending upon the relative magnitudes of L, R, G, and C),

$$\gamma_{\frac{1}{2}} = \pm[(R + i\omega L)(G + i\omega C)]^{1/2} = \pm [(RG - \omega^2 LC) + (RC + LG)i\omega]^{1/2} \tag{1.5g''}$$

Now, γ_1 and γ_2 are complex, i.e., generally,

$$\gamma_{\frac{1}{2}} = \alpha_{\frac{1}{2}} + i\beta_{\frac{1}{2}} \tag{1.11}$$

However, for the last three of the above cases, i.e., the nonflow ones,

$$\gamma_1 = -\gamma_2$$

so the solutions to Equations (1.5e′), (1.5f′), and (1.5g′) are all of the form

$$V = Ae^{-\gamma z} + Be^{+\gamma z} \tag{1.12}$$

where

$$\gamma = \alpha + i\beta \tag{1.13}$$

α and β both being positive. In contrast, the solutions to Equations (1.5a′)–(1.5d′) are much different and form the subject matter of a large part of this book.

To return to the other three cases illustrated, it is seen that: For the example of a lossy system, from Equations (1.5e″) and (1.11),

$$\alpha = \beta = (\omega/2a)^{1/2}$$

For the lossless system from Equations (1.5f′) and (1.11),

$$\alpha = 0, \qquad \beta = \omega/c$$

For the transmission line, $\alpha \neq \beta$ (generally), and as stated, algebraic expressions for α and β in this last case may be found from Equations (1.5g″) and (1.11).

1.4 INCIDENT AND REFLECTED WAVES

The solution to the second-order wave equation in the steady cyclic state [Equation (1.12)] is made up of two parts, containing $\pm\gamma$ and two constants, A and B. Values of the latter can be put in so that the wave will comply with certain stipulations, i.e., that boundary conditions are satisfied. The result would be that the amplitude and phase angle would be specified as functions of distance, but these functions would be inconvenient to use—far less so than if the system were infinitely long.

1.4.1 The Infinite System

In the propagation constant $\gamma = \alpha + i\beta$ the first term α describes how the absolute value of the amplitude alters with distance, the second term β describes how the phase angle alters with distance. That is to say, γ describes the phase vector completely. For this reason it is called the propagation constant, while α is called the *attenuation constant* and β the *phase constant.*

For an *infinite system*, B must be zero, for otherwise V would be infinite as $z \longrightarrow \infty$, since α and β are positive. Hence, for an infinite system

$$V = Ae^{-\gamma z} = Ae^{-\alpha z}e^{-i\beta z} \tag{1.14}$$

and so

$$v = Ve^{i\omega t} = Ae^{-\alpha z}e^{i(\omega t - \beta z)} \tag{1.15}$$

Now, as stated, Equation (1.15) will give the instantaneous value of v at any time and any distance z if the real (or imaginary) trigonometric part is taken. That is, $v(t, z)$ is the projection of a phasor whose length is $Ae^{-\alpha z}$ and which makes an angle $(\omega t - \beta z)$ with some reference angle at $t = z = 0$. This angle repeats itself (i.e., the phasor rotates through 2π radians) at constant t when $\beta\lambda = 2\pi$; i.e.,

$$\lambda = 2\pi/\beta \tag{1.16}$$

Furthermore, the phasor can be considered to have moved through this distance λ in the time required for one complete revolution, for, by §1.2, the velocity of travel of this phasor, i.e., the wave velocity $V_w = \lambda/\tilde{t}$, hence

$$\tilde{t} = \frac{\lambda(\text{cm})}{V_w(\text{cm sec}^{-1})} = \frac{2\pi(\text{radians})}{\omega\,(\text{radians sec}^{-1})} \tag{1.17}$$

Also, by Equation (1.16),

$$V_w = \omega/\beta \tag{1.18}$$

Figure 1.7(a) shows the position of the phasor at different fractions of λ

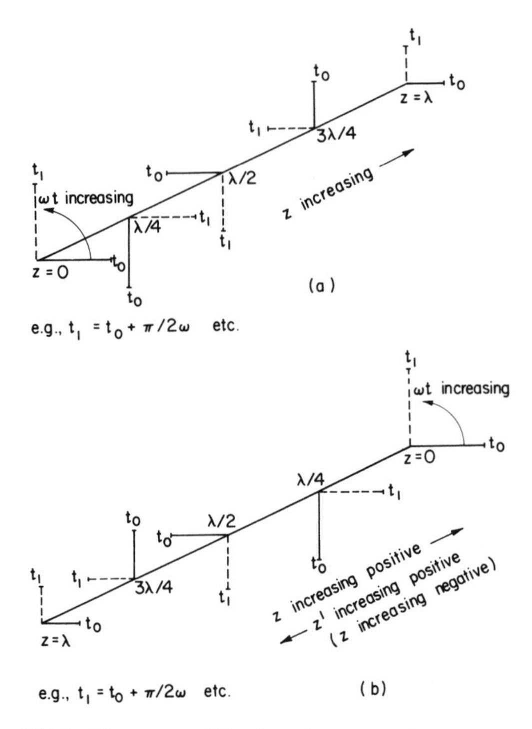

Fig. 1.7. Phasors of (a) incident wave; (b) reflected wave. Both are given for the same times (e.g., t_0 and t_1) and at the same values of z.

at constant time, and at various fractions of $\tilde{t}$ at constant z; e.g., $t_1 = t_0 + \frac{1}{4}\tilde{t}$.

Two additional points may be made. One is that the expression for a wave, viz.,

$$v = A\cos(V_w t - z)$$

(for example) contains the trigonometric function of a dimensioned quantity, viz., length, so it is desirable to write it in dimensionless form, e.g.,

$$v = A\cos\{2\pi[(t/\tilde{t}) - (z/\lambda)]\} \tag{1.19a}$$

$$= A\cos[(2\pi/\lambda)(V_w t - z)] \tag{1.19b}$$

$$= A\cos[\omega t - (2\pi z/\lambda)] \tag{1.19c}$$

$$= A\cos(\omega t - \psi) \tag{1.19d}$$

$$= A\cos\{\omega[t - (\psi/\omega)]\} \tag{1.19e}$$

$$= A\cos[\omega(t - t_{\text{lag}})] \tag{1.19f}$$

$$= A\cos\{\omega[t - (z/V_w)]\} \tag{1.19g}$$

Of these equations, (1.19b) stresses the idea of a wave of wavelength λ and velocity V_w, while Equation (1.19d) brings in the idea of a phase angle and hence of the generation of a periodic oscillation at fixed z by a rotating vector, and Equation (1.19f) points out that the phase difference between a wave and a reference may be expressed in terms of a time lag. All these three concepts have applications in appropriate contexts.

The second point is that trigonometric expansion of any of these equations will show that at any given value of z the resultant oscillation (with time) may be thought of as the sum of two oscillations of the same frequency, but with a constant phase-angle difference of $\pi/2$ and of amplitudes that are functions of z. For example, such an expansion of Equation (1.19b) would give

$$v = \{[A \cos(2\pi z/\lambda)] \cos(2\pi V_w/\lambda)t + [A \sin(2\pi z/\lambda)] \sin(2\pi V_w/\lambda)t\} \quad (1.20)$$

1.4.2 Noninfinite Systems (Steady Cyclic State)

When the system is not infinite the consequences are, as stated, that for a steady cyclic state, B in Equation (1.12) is not zero and the simple form of the equations of variation of amplitude and phase angle with distance is lost. However (and again for the case of the steady cyclic state), two points can be made:

(a) Experimentally, it is relatively easy to regain the simplicity of the infinite bed in a rigorous way.
(b) The analysis of the finite system, when desired, is made easier by adopting the concepts of other branches of traveling wave engineering (particularly the study of long transmission lines) by introducing the idea of reflected waves.

Both (a) and (b) will be discussed in Chapter 6, but by way of introduction to the idea of reflections, it will be seen that the term $B \exp(\gamma z)$ in Equation (1.12) could be written as $B \exp(-\gamma z')$, where z' is a distance increasing in the opposite direction to z. Thus, from Equation (1.12) the value of v could be written as

$$v = Ve^{i\omega t} \quad (1.21)$$

where

$$V = \boldsymbol{v}^{+} + \boldsymbol{v}^{-} \quad \text{(vector sum)} \quad (1.22)$$

with

$$\boldsymbol{v}^{+} = A \exp(-\gamma z) \quad (1.23)$$

$$\boldsymbol{v}^{-} = B \exp(-\gamma z') \quad (1.24)$$

Thus, v^+ would be the value at t and z of a wave whose phasor was specified in direction by $\exp[i(\omega t - \beta z)]$ and v^- would be the corresponding value of a wave whose phasor was specified in direction by $\exp[i(\omega t + \beta z)]$ or $\exp[i(\omega t - \beta z')]$.

Hence, if Figure 1.7(a) represented the phasor at t and z of the wave traveling in the direction of z increasing positive, then Figure 1.7(b) would represent the phasor of a wave traveling in the direction of z increasing negative. The directions of rotation should be noted; if both diagrams are viewed from the left-hand end looking from left to right, then an increasing positive value of ωt results in the phasor rotating in an anticlockwise direction (with the convention adopted here). So, in Figure 1.7(a) as z increases (i.e., the observer travels toward the right), the angle of the phasor, viz., $(\omega t - \beta z)$, has increased in a negative (i.e., clockwise) direction at constant t. On the other hand, in Figure 1.7(b), if the same procedure is followed, the angle of the phasor, viz., $(\omega t - \beta z') = (\omega t + \beta z)$, will have increased in a positive (anticlockwise) direction at constant t.

Since the terms clockwise and anticlockwise are not absolute (to the observer they will represent different directions, depending on whether he looks at the front or the back of the clock), it is better to describe the rotation of the phasors as being that of a right-handed screw thread in Figure 1.7(a) and that of a left-handed screw thread in Figure 1.7(b). One other point of difference exists between Figures 1.7(a) and 1.7(b). If the wave amplitude is a function of distance (i.e., $\alpha \neq 0$), then the length of the phasor diminishes; in Figure 1.7(a) it decreases from left to right (z increasing positive), while in Figure 1.7(b) it decreases from right to left (z' increasing positive).

1.4.3 Noninfinite Systems (Pulses)

When the wave is not a repetitive one, the boundary conditions at the entrance and exit of the bed have again to be satisfied, of course. The idea of reflections could still be used, but in general, the only analyses that appear to have been published apply to the case where there is plug flow of the medium, the disturbance is an impulse or a pulse, but there is no reservoir term q_j. A good general review will be found in Levenspiel and Bischoff [5] and the subject will be discussed in Chapter 5.

1.5 LOSSY AND LOSSLESS SYSTEMS

Equations (1.5e′) and (1.5f′) were said to describe "lossy" and "lossless" systems, respectively; the first appellation is only relative, of course. The terms arise because the resulting equations for γ, viz., (1.5e″) and (1.5f″) combined with Equation (1.11), give

$$\alpha = \beta = (\omega/2a)^{1/2} = \omega'$$

and

$$\alpha = 0, \qquad \beta = \omega/c$$

respectively. So, in the first system the amplitude is attenuated markedly; for, in one wavelength ($z = \lambda$)

$$|V|(z = \lambda)/|V|(z = 0) = e^{-\alpha\lambda} = e^{-\alpha(2\pi/\beta)} = e^{-2\pi}$$

which is very large. Note that the attenuation per wavelength is used, so that the words "attenuation" and "lossy" may be comparable for different systems between which wavelengths may vary widely. A lossy system is thus one in which the attenuation per wavelength is large.

In the second system the wave is not attenuated at all, because $\alpha = 0$. In general, the attenuation of a wave is measured by the magnitude of the ratio α/β.

The attenuation in the systems described by Equation (1.5e′), and others, may take one of a range of values, depending upon the magnitude of the appropriate quantities. As for Equations (1.5a′)–(1.5c′), these form the subject of an appreciable part of this work. The attenuation may be large or small, while in the solution—of the form of Equation (1.10a)—γ_1 and γ_2 are not numerically equal. This is an interesting and highly significant fact.

1.6 NONSINUSOIDAL WAVES

One of the advantages of a sinusoidal wave (listed in Chapter 9) is that it does not change shape as it passes through a linear system. This follows from Equations (1.17) and (1.18), which demonstrate that the wavelength and velocity are constant. The amplitude may decrease as the wave progresses, but the ideas of a velocity and a wavelength are still valid if the distance between corresponding points (such as peaks or intersection with the mean line) is taken. For, if the wave equation is written $v = Ae^{-\alpha z}\sin(\omega t - \beta z)$, i.e., by taking the imaginary part of Equation (1.14), then at the midpoint of the swing the above statement is true if:

(i) $v = 0$ at $Ae^{0}\sin\omega t = 0$

(ii) $v = 0$ at $Ae^{-\alpha\lambda}\sin(\omega t - \beta\lambda) = 0$

and if, at the extremes of the swing,

(iii) $(\partial v/\partial z)_t = 0$ at $A[(-\alpha)e^{0}\sin\omega t + (-\beta)e^{0}\cos\omega t] = 0$

(iv) $(\partial v/\partial z)_t = 0$

$$\text{at} \quad A[(-\alpha)e^{-\alpha\lambda}\sin(\omega t - \beta\lambda) + (-\beta)e^{-\alpha\lambda}\cos(\omega t - \beta\lambda)] = 0$$

Now, since $\beta\lambda = 2\pi$, it follows that $v = 0$ at both $z = 0$ and $z = \lambda$, and that $(\partial v/\partial z)_t = 0$ at both $z = 0$ and $z = \lambda$ (a different origin from the previous one being used). Hence, the statement is true and is illustrated in Figure 1.8.

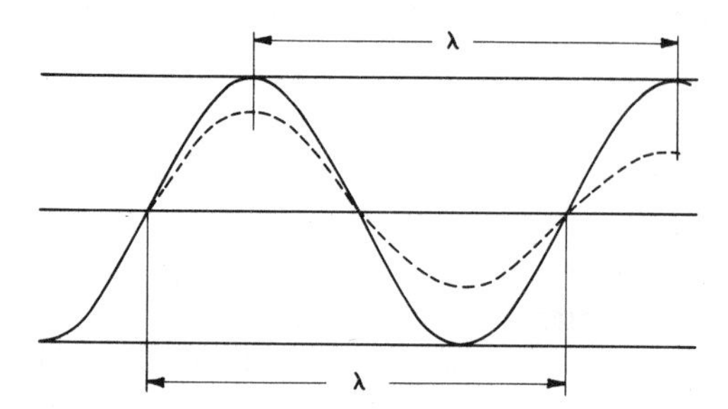

Fig. 1.8. Wavelength has the same meaning for constant and attenuated pure sine waves.

However, if the wave is not a pure sine wave, it may be considered to be the resultant of two or more pure sine waves of different frequencies, i.e., of harmonics; the frequencies being different, it follows that the harmonics travel at different velocities [Equation (1.17)] and so the wave changes shape as z increases. Thus, if, say, a pulse is put into a flowing medium, its possible shapes at different values of z might be as in Figure 1.9.

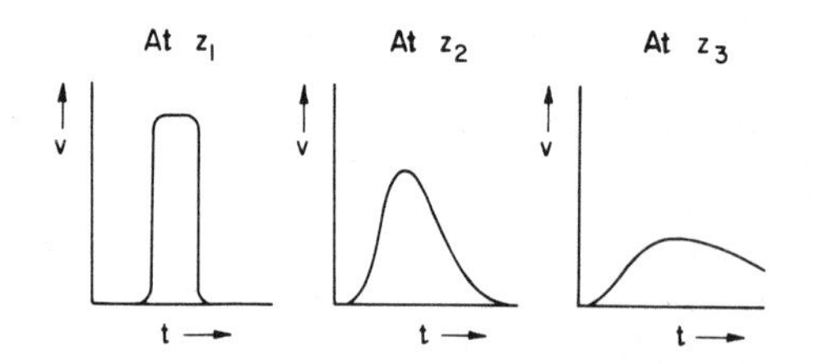

Fig. 1.9. Nonsinusoidal pulse or wave; change of shape in a dispersive medium.

If there were repetitious pulses, the concepts of wavelength and velocity are not possible along the above lines, since there are no corresponding points as clearly discernible as midpoints or peaks. In fact, the use of the peak, i.e., maximum of such a wave, leads to incorrect deductions, as will be shown.

Such a pulse, as is shown in Figure 1.9, could be described in terms of its Fourier transform into harmonic components or in terms of its moments along the lines of a probability distribution. The first moment, in fact, can be used to define a velocity, as will be shown. These aspects will be treated in Chapter 4. The behavior of each harmonic would be as a pure sine wave of the same frequency, and the portion of the book devoted to periodic waves need be confined only to pure sine waves for this reason. On the other hand, the full description of a pulse as in Figure 1.9 needs a large number of moments, which are difficult to measure accurately, the higher ones having tedious algebraic expressions, and the imposition of a finite bed length introduces overwhelming manipulatory difficulties for any but the simplest cases.

1.7 STANDING WAVES

Standing (or stationary) waves, by definition, do not travel, i.e., instantaneous snapshots of some pictorial representation of the wave at successive intervals of time are not displaced repetitions. Their physical effect and mathematical representation are thus different from traveling (or progressive) waves. For most systems they are important, and are discussed in texts on waves and acoustics (see Moore [3]).

They may be considered to be the net result of two waves of the same amplitude and frequency, but traveling in opposite directions. Thus, by applying Equation (1.20) to both positive and negative waves [e.g., in the form of Equation (1.19b)] there arise (provided that at $z = 0$ the waves are in phase):

$$v^+ = \{[A\cos(2\pi z/\lambda)\cos(2\pi V_w/\lambda)t] + [A\sin(2\pi z/\lambda)\sin(2\pi V_w/\lambda)t]\}$$

$$v^- = \{[A\cos(2\pi z/\lambda)\cos(2\pi V_w/\lambda)t] - [A\sin(2\pi z/\lambda)\sin(2\pi V_w/\lambda)t]\}$$

Hence, the resultant $v = v^+ + v^-$ is

$$v = 2A\cos(2\pi z/\lambda)\cos(2\pi V_w/\lambda)t$$

This gives a sinusoidal variation of v either with z at fixed t, or with t at fixed z. Further, when

$$2\pi z/\lambda = 0, \pi, 2\pi, \ldots$$

the amplitude of the variation with time is a maximum (equal to twice the amplitude of either wave). Again, when

$$2\pi z/\lambda = \pi/2, 3\pi/2, 5\pi/2, \ldots$$

the amplitude is zero for all values of t. The former positions are called *antinodes*, and the latter are called *nodes*. It will be seen that the distance between antinode and antinode, and between node and node, is $\lambda/2$. Also, the distance between antinode and the next node is $\lambda/4$.

Since the physical beginning and end of the system must also satisfy the condition of there being either a node or an antinode there (depending on circumstances), it follows that the length of the system will be determined by the particular frequency used.

In a flowing medium, with longitudinal boundaries a finite distance apart, it will be found in Chapter 6 that positive and negative waves appear. However, in contrast to the preceding situation, they are unlikely to result in standing waves, for the positive and negative waves are not equal in amplitude and phase even at a discontinuity (i.e., "reflecting surface") and furthermore, both waves usually decrease in amplitude in the direction of travel, the negative wave very markedly so. It is left as an exercise to apply the argument of the beginning of this section to such a situation. It will be found that the variation at fixed z is still sinusoidal with time but the amplitude is not a simple harmonic function of distance.

REFERENCES

[1] M. J. Lighthill, and G. B. Whitham, On Kinematic Waves I. Flood Movement in Long Rivers. *Proc. Roy. Soc.* (*London*) **229A**, 281–316 (1955).
[2] O. Levenspiel, "Chemical Reaction Engineering." Wiley, New York, 1971.
[3] R. K. Moore, "Traveling Wave Engineering." McGraw-Hill, New York, 1960.
[4] C. A. Coulson, "Waves." Oliver & Boyd, Edinburgh, 1947.
[5] O. Levenspiel and K. B. Bischoff, Patterns of Flow in Chemical Process Vessels. "Advances in Chemical Engineering," Vol. 4. Academic Press, New York, 1963.

Chapter 2

Sources in One-Dimensional Flow

2.1 INTRODUCTION

This chapter describes the effect of introducing either an impulse or a steady source into a stream of uniform velocity. This source may be at a point, along a line, or over an area; in the first two cases the concentration that results will be a function of three spatial dimensions, and although not widely used experimentally, the consideration of point and line sources leads logically to the frequently employed plane source. The equations in this chapter that describe the concentrations assume that the dispersion coefficient D is the same in all directions and that the system is infinite in extent; i.e., there are no boundary conditions (except at $\pm\infty$) to be satisfied.

If a point or line source were to be used in practice, it would be necessary to consider both these assumptions, for any practical system is physically bounded and in the case of, say, flow in a packed bed, D is not the same in directions parallel to and normal to the flow direction; see, for example, Levenspiel and Bischoff [1].

2.2 SOURCES

An idealization that has proved to be of great use ever since its introduction by Kelvin is that of an *instantaneous source*; it is that a finite amount of heat or matter or electricity or anything else is released—or irrupts—in a vanishingly small interval of time and thereafter its history is molded by the effects that its surroundings have upon it. It is probably safe to say that to make one experimentally is impossible; even a close approximation is difficult, for the closer an irruption resembles the idealization, the greater will be the tendency for the pulse to have effects not desired in the system that the source is to probe. For example, a truly instantaneous heat source would have disruptive consequences—local vaporization in a liquid or solid, and shock waves in any system. Nevertheless, experimental approximations are of value in finding out information about a system (an idea of the effect of the nonideality of a source is contained in Carslaw and Jaeger [2]), while the theoretical concept is used in analyzing the effect upon a system of a continuous source: this is defined as a continuous series of instantaneous sources; that is to say, this source lasts for a period of time that may be finite or infinite, but it is not infinitesimal.

An account of these instantaneous and continuous sources will be found in Chapter 10 of Carslaw and Jaeger [2]. In the present account the symbol $v(x, y, z, t)$, denoting concentration, will have two subscripts taken from Table 2.1, which applies to this chapter only. Note that the source will be expressed as an amount Q, *or* amount per unit length, *or* amount per unit area, depending an circumstances.

(*i*) *An Instantaneous Point Source in a Stationary Medium: Unsteady State*

An instantaneous point source of *strength* Q releases the finite quantity of heat or matter Q in a vanishingly small interval of time in a vanishingly small domain in space at time $t = 0$. The resulting concentration at any point (x, y, z) in the surrounding medium at time t will be denoted by $v_1(x, y, z, t)$; it is convenient to make the origin of the space variables the position of the source. An equation for v_1 is given as Equation (2.8).

TABLE 2.1
SOURCES: MEANING OF SYMBOLS AND SUBSCRIPTS

First subscript	Source	Medium	State
1	Instantaneous	Stationary	Unsteady[a]
2	Continuous	Stationary	Unsteady[a]
3	Instantaneous	Moving	Unsteady[a]
4	Continuous	Moving	Unsteady
5	Continuous	Moving	Steady

Second subscript	Meaning
pt	Point
l	Line
pl	Plane

Example	Meaning
$(v_1)_{pt}$	Concentration arising from an instantaneous point source in a moving medium
$(v_4)_{pl}$	Concentration arising from a continuous plane source in a moving medium, unsteady state

Sources	Type
Q	Instantaneous
q	Continuous

Example	Meaning	Dimensions
Q_{pt}	Instantaneous point source	Q
Q_l	Instantaneous line source	QL^{-1}
Q_{pl}	Instantaneous plane source	QL^{-2}
q_{pt}	Steady point source	QT^{-1}
q_l	Steady line source	$QT^{-1}L^{-1}$
q_{pl}	Steady plane source	$QT^{-1}L^{-2}$

[a]There is, of course, no steady state corresponding to any of these situations.

(*ii*) *Instantaneous Line and Surface Sourses in a Stationary Medium: Unsteady State*

The concentration distribution due to a source that is of finite amount in an infinitesimal time, but is spread along a line or over a surface (rather than concentrated at a point), may be derived by considering that there are an infinity of point sources spread along the line or over the surface, all irrupting simultaneously. The concentration function can be obtained by integration of v_1 along the line or over the surface. For example, the concentration due to *a line source distributed from* $-\infty$ *to* $+\infty$ *along the x axis* could be obtained as

$$(v_1)_l = \int_{-\infty}^{+\infty} (v_1)_{pt}\, dx \tag{2.1}$$

while the concentration due to *an infinite plane source distributed in the XY plane* could be obtained by either

$$(v_1)_{pl} = \int_{-\infty}^{\infty} (v_1)_l\, dy \tag{2.2}$$

or

$$(v_1)_{pl} = \int_{-\infty}^{\infty}\int_{-\infty}^{\infty} (v_1)_{pt}\, dx\, dy \tag{2.3}$$

(see Carslaw and Jaeger [2]).

(iii) A Continuous Source in a Stationary Medium: Unsteady State

In contrast to an instantaneous source, a continuous source which has emitted for a finite time t(i.e., t is zero at the commencement of emission) may be regarded as a continuous succession of instantaneous sources. The concentration distribution due to this can also be obtained by integration of that due to an instantaneous source but with respect to *time*; moreover, the argument is a little more subtle.

Consider the amount of heat or matter emitted in the period from t' to $t' + dt'$. (The origin of t' is not important.) This amount will be $q(t')\, dt'$, where $q(t')$ is the mean rate of evolution over the interval between t' and $t' + dt'$ and can be considered as an instantaneous source at time t'. The effect at time t—that is, at $(t - t')$ seconds later—caused by this elemental source, where t is the instant that we are considering, is obtained by substituting $(t - t')$ for t and $q(t')\, dt'$ for q in the equation for v_1. The gross effect of all the elemental sources is obtained by summing the contributions of the individual ones from $t' = 0$ to $t' = t$. *Hence, for a continuous source in a stationary medium:*

$$v_2 = \int_0^t v_1(x, y, z, (t - t'))\, dt' \tag{2.4}$$

generally.

The spatial integrations of (ii) and the time integrations of (iii) can be combined. Thus (for example), for the continuous source covering the XY plane in a defined area, and which has been emitting for time t, the concentration will be given by

$$\begin{aligned}(v_2)_{pl} &= \int_0^t (v_1)_{pl}(x, y, z, (t - t'))\, dt' \\ &= \int^y\int^x\int_0^t (v_1)_{pt}(x, y, z, (t - t'))\, dt\, dx\, dy \end{aligned} \tag{2.5}$$

(iv) *A Continuous Source in a Moving Medium: Unsteady State*

If the continuous source exists at a fixed point in space but in a medium that moves in the z direction, then the argument has again to be extended a little.

Consider an observer at point $A(x, y, z)$: the portion of the fluid—of vanishingly small size—that surrounds A at time t *was* at point B, a distance $U(t - t')$ upstream at an earlier time t'. Over the interval t' to $t' + dt'$ the source—which at that instant was a distance $z - U(t - t')$ away from the parcel of fluid at B—gave out the amount of heat $q(t')\,dt'$ as an instantaneous source that started to drift downstream at the instant of emission. Since, however, all other parts of the fluid are drifting downstream likewise, it follows that relative distances are preserved: in particular, as the fluid that was at point B (and is now referred to as B) drifts downstream, it will be influenced by this degenerating source as it, too, drifts downstream at a distance $z - U(t - t')$ away. So the concentration of fluid B will be governed by the equation that describes how the concentration at a point varies with time due to diffusion from an instantaneous source at a fixed distance away in a stationary medium. Thus, at some time t'' (subsequent to t') the concentration of B will be $v_1(x, y, [z - U(t - t')], (t - t'))$ in the nomenclature of this section. But t' was any time in the interval from zero (when the source started to emit) to t (the instant that we are considering), and so the concentration of the parcel of fluid B will be governed not only by the effect that the instantaneous source that irrupted at t' has on it, but also by the effects due to every other elemental source that irrupts from $t = 0$ to $t = t$. Since the concentration is an additive quantity in this case, the final concentration will be the sum of the effects due to all those other elemental sources that have irrupted, each at a different distance away from B and at a different time. Thus, for dt' made vanishingly small the concentration of the parcel of fluid B when it surrounds point A (i.e., when $t'' = t$) will be the integral of all these effects, and so it can be expressed as

$$v_4(z, y, z, t) = \int_{t'=0}^{t'=t} v_1\big(x, y, [z - U(t - t')], (t - t')\big)\,dt' \qquad (2.6)$$

q, U, and D being parameters in v_4.

(v) *A Continuous Source in a Moving Medium: Steady State*

In a moving medium there is a continual removal of heat or matter by the renewal, by the stream, of the fluid that is receiving it from the source. If this source were to continue emitting for a long time, ultimately a state of equilibrium would be reached. The resulting concentration distribution in space would be obtained by making the limit t tend to infinity in the appropriate integration in Equation (2.6).

So,

$$v_5(z, y, z) = \left\{\int_0^t v_1(x, y, [z - U(t - t')], (t - t'))\, dt'\right\}_{t\to\infty} \tag{2.7}$$

2.3 SOME EXAMPLES OF SOURCES IN A MOVING MEDIUM

Some examples of the foregoing arguments in Section 2.2 will now be given for the following cases:

Subsection	Source	State
2.3.1	Point, continuous	Unsteady
2.3.2	Point, continuous	Steady
2.3.3	Plane, continuous	Steady
2.3.4	Plane, continuous	Unsteady
2.3.5	Plane, instantaneous	—

2.3.1 A Continuous Point Source in a Moving Medium: Unsteady State

The concentration at any place and time due to an instantaneous point source of strength Q_{pt} at the origin of coordinates in an isotropic infinite stationary medium is, from Carslaw and Jaeger [2]

$$(v_1)_{pt} = [Q/8(\pi Dt)^{3/2}]\exp[-(x^2 + y^2 + z^2)/4Dt] \tag{2.8}$$

satisfying both the differential equation

$$\frac{\partial^2 v}{\partial x^2} + \frac{\partial^2 v}{\partial y^2} + \frac{\partial^2 v}{\partial z^2} - \frac{1}{D}\frac{\partial v}{\partial t} = 0 \tag{2.9}$$

as well as the condition that v tends to zero when t tends to zero (i.e., to the instant that the source irrupts) for all points except at the origin, where it tends to infinity.

A continuous point source of constant strength q_{pt} fixed in space in a medium that moves at velocity U along the z axis will give a resulting concentration distribution [by the reasoning that led to Equation (2.6)] as:

$$(v_4)_{pt} = [q_{pt}/8(\pi D)^{3/2}] \times \int_0^t (t - t')^{-3/2} \exp -(\{x^2 + y^2 + [z - U(t - t')]^2\}/4D(t - t'))\, dt' \tag{2.10}$$

Appendix 3 deals with the problem of evaluating such integrals and the analytic expression resulting from the method described there is obtained, after a slight rearrangement of Equation (2.10), to give

$$(v_4)_{pt} = \{q_{pt}[\exp(zU/2D)]/8(\pi D)^{3/2}\} \times \int_0^t (t - t')^{-3/2} \exp -\{[R^2/4D(t - t')] + [U^2(t - t')/4D]\}\, dt' \quad (2.11)$$

where $R^2 = x^2 + y^2 + z^2$, as

$$(v_4)_{pt} = \{q_{pt}[\exp(zU/2D)]/8\pi DR\} \times \big(2\cosh(RU/2D) + [\exp(-RU/2D)]\,\mathrm{erf}\{(Ut^{1/2}/2D^{1/2}) - [R/2(Dt)^{1/2}]\} - [\exp(RU/2D)]\,\mathrm{erf}\{(Ut^{1/2}/2D^{1/2}) + [R/2(Dt)^{1/2}]\}\big) \quad (2.12)$$

which is therefore the solution to the differential equation describing the system, viz.,

$$\frac{\partial^2 v}{\partial x^2} + \frac{\partial^2 v}{\partial y^2} + \frac{\partial^2 v}{\partial z^2} - \frac{U}{D}\frac{\partial v}{\partial z} - \frac{1}{D}\frac{\partial v}{\partial t} = 0 \quad (2.13)$$

subject to the conditions

$$\begin{array}{llll} v \longrightarrow \infty, & R = 0, & t \longrightarrow 0 & \\ v = 0, & R \neq 0, & t = 0 & \\ v \text{ is finite}, & R = 0, & t \neq 0 & \\ v \text{ is finite}, & R \longrightarrow \pm\infty & \text{for all } t & \end{array} \quad (2.14)$$

2.3.2 A Continuous Point Source in a Moving Medium: Steady State

The argument that led to Equation (2.7) gives, for this specific case

$$(v_5)_{pt} = \int_0^\infty (v_1)_{pt}\big(x, y, [z - U(t - t')], (t - t')\big)\, dt' \quad (2.15)$$

and so the limit $t \longrightarrow \infty$ in Equation (2.12) leads to

$$(v_5)_{pt} = [q_{pt}/4\pi D(z^2 + r^2)^{1/2}] \exp\{-U[(r^2 + z^2)^{1/2} - z]/2D\} \quad (2.16)$$

where $r^2 = x^2 + y^2$, as shown in Appendix 3 (Wilson [3]).

This is an example of steady-state radial diffusion. The concentration v does not become zero at any finite value of r or z, even when they are large and negative, provided that $D \neq 0$; i.e., the effect of this source is felt an infinite distance upstream.

2.3.3 A Continuous Plane Source in a Moving Medium: Steady State

For this case an equation for $(v_5)_{pl}$ may be obtained in either of two ways: (a) by using the idea that a steady state is the limiting case after infinite time

[Equation (2.7)] or (b) by the argument that states that the effect of a plane source is the sum of an infinity of point sources [Equation (2.3)]. Thus, the substitution of Equation (2.8) (with Q replaced by the source q_{pl}, which is a rate of input per unit area) for $(v_1)_{pt}$ gives

$$(v_5)_{pl} = [q_{pl}/8(\pi D)^{3/2}] \int_{-\infty}^{\infty} \int_{-\infty}^{\infty} \int_{0}^{\infty} (t - t')^{-3/2} \times \exp\big(-\{x^2 + y^2 + [z - U(t - t')]^2\}/4D(t - t')\big)\, dt'\, dx\, dy \tag{2.17}$$

The substitution $\xi^2 = x^2/4D(t - t')$ and subsequent integration with respect to y, followed by the substitution $\xi^2 = y^2/4D(t - t')$ and subsequent integration with respect to x, leads, via the form $\int_{-\infty}^{\infty} \exp(-\xi^2)\, d\xi = \pi^{1/2}$ (see Gibson [4]) in each case, to

$$(v_5)_{pl} = [q_{pl}/2\pi D)^{1/2}] \int_{0}^{\infty} (t - t')^{-1/2} \times \exp -\{[z - U(t - t')]^2/[4D(t - t')]\}\, dt' \tag{2.18}$$

Since z can be positive or negative, being measured upstream or downstream from the source (the origin), this equation can be written

$$(v_5)_{pl} = q_{pl}[\exp (Uz/2D)]/2(\pi D)^{1/2}(I_{(-1/2)\infty}) \tag{2.19}$$

in the nomenclature of Appendix 3, where it is shown that

$$I_{(-1/2)\infty} \left\{ \equiv \int_{0}^{\infty} \xi^{-1/2} \exp -[(a^2/\xi) + b^2\xi] d\xi \right\} = (\pi^{1/2}/b) \exp(-2ab)$$

(a and b being positive) and so

$$(v_5)_{pl} = q_{pl}[\exp \pm(Uz/2D)]/\{[2(\pi D)^{1/2}]\}\{[2(\pi D)^{1/2}/U] \exp(-Uz/2D)\}$$

where the positive sign is to be taken when the point lies downstream of the source and vice versa; i.e.,

$$(v_5)_{pl} = q_{pl}/U \qquad \text{if } z \text{ is measured downstream} \tag{2.20a}$$

$$= (q_{pl}/U) \exp(-Uz/D) \qquad \text{when } z \text{ is taken as positive in the upstream direction} \tag{2.20b}$$

The above logic is to show how the effect of a plane, continuous source can be deduced from the properties of an instantaneous point source: the last results are, in fact, readily obtainable directly from the differential equation describing the system (Wilson [3]), which simplifies to the steady-state one in the z direction, viz.,

$$D(d^2v/dz^2) - U(dv/dz) = 0$$

subject to the condition that

$$v = 0, \qquad z \longrightarrow -\infty$$

The solutions (2.20a) and (2.20b) readily follow. (The solution for points downstream of the source could be written down from first principles.)

The Use of Equations (2.20a) and (2.20b)

It follows from Equation (2.20b) that if a point is found upstream of the source where the concentration is $1/e$ times that downstream of the source, and the distance between this point and the source—*a characteristic length*—being z_D, then z_D will be numerically equal to D/U. If, further, U is known, then D can be found.

Again, the value of the plug-flow velocity U could be found by measuring the difference between the steady-state downstream concentration and that obtaining sufficiently far upstream, for then, of course, $U = q_{\text{pl}}/(v_5)_{\text{pl}}$. (This implies that the concentration is uniform across the flow; either the source is a true plane one or because radial diffusion has made uniform the concentration arising from, say, a point source.) This is the basis of the calorimetric mass flow-meter, which can be an accurate absolute method of measuring fluid flows, and of the salt-dilution or tracer dilution method, which is used to measure large liquid flow rates. In both a steady, known input is used.

2.3.4 A Continuous Plane Source in a Moving Medium: Unsteady State

(a) The Use of the Concept of a Continuous Source

An equation for the concentration (as a function of space and time) is derived in the same way as for a steady state, but the integral with respect to time has a finite upper limit t; this may mean that deriving an analytic expression for the concentration may be more difficult. For the present case we would have [as Equation (2.18) but with the limit of time not infinite],

$$(v_4)_{\text{pl}} = [q_{\text{pl}}/2(\pi D)^{1/2}][\exp(Uz/2D)] \int_0^t (t - t')^{-1/2} \times \exp -\{[z^2/4D(t - t')] + [U^2(t - t')/4D]\}\, dt' \tag{2.21}$$

The solution to this, obtained in Appendix 3, is

$$(v_4)_{\text{pl}} = [q_{\text{pl}}/2U) \exp(Uz/2D)]([\exp(-Uz/2D)] \operatorname{erf}\{(Ut^{1/2}/2D^{1/2}) - [z/2(Dt)^{1/2}]\} + [\exp(Uz/2D)] \operatorname{erf}\{(Ut^{1/2}/2D^{1/2}) + [z/2(Dt)^{1/2}]\} - 2\sinh(Uz/2D)) \tag{2.22}$$

if $z \geq 0$.

For $z < 0$, only the first term $\exp(Uz/2D)$ has the sign of z altered, because the value of the integral in Equation (2.21) is the same for positive and negative values of z. So,

$$\begin{aligned}(v_4)_{pl} = {} & [(q_{pl}/2U)\exp(-Uz/2D)](\exp(-Uz/2D)\operatorname{erf}\{(Ut^{1/2}/2D^{1/2})\\ & - [z/2(Dt)^{1/2}]\} + [\exp(Uz/2D)]\operatorname{erf}\{(Ut^{1/2}/2D^{1/2})\\ & + [z/2(Dt)^{1/2}]\} - 2\sinh(Uz/2D))\end{aligned} \tag{2.23}$$

if $z < 0$, the absolute value $|z|$ being used in computations involving Equations (2.22) and (2.23).

(*b*) *The Direct Solution of the Differential Equation*

Equation (2.21) can also be arrived at by a direct solution of the differential equation governing this case, which is illustrated in Figure 2.1.

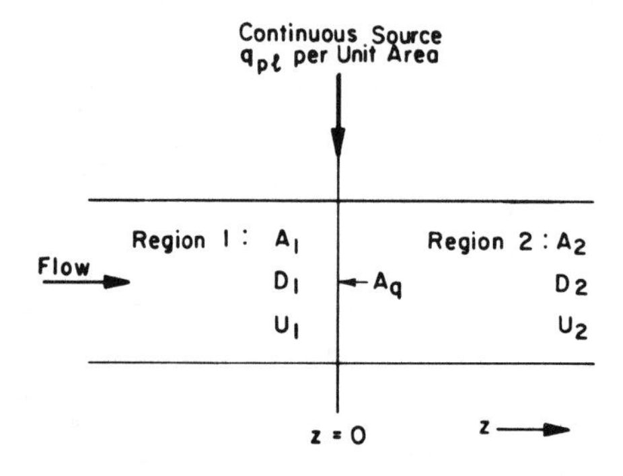

Fig. 2.1. Plane source in one-dimensional flow.

The source is at $z = 0$, which divides region 1 from region 2: values of the parameters U and D will be identified by corresponding subscripts.

The equation applying to either region, away from $z = 0$, is

$$\frac{\partial^2 v}{\partial z^2} - \frac{U}{D}\frac{\partial v}{\partial z} - \frac{1}{D}\frac{\partial v}{\partial t} = 0 \tag{2.24}$$

with conditions

$$v \longrightarrow 0 \quad \text{for} \quad t \longrightarrow 0 \quad \text{for} \quad z \neq 0 \tag{2.25a}$$

$$v \quad \text{is finite for} \quad z \longrightarrow \pm\infty \quad \text{for all} \quad t \tag{2.25b}$$

$$A_1\rho_1U_1 = A_2\rho_2U_2 \tag{2.25c}$$

$$A_1(U_1v_1 - D_1\,\partial v_1/\partial z)_{z\to 0_-} = A_2(U_2v_2 - D_2\,\partial v_2/\partial z)_{z\to 0_+} - A_qq_{pl} \tag{2.26}$$

(from a balance across the source plane; 0_- and 0_+ are used to signify the approach of z to zero at the source plane from the regions where z is less than, and greater than zero, respectively)

$$v_1 = v_2 \qquad \text{at} \quad z = 0 \tag{2.27}$$

(See, however, Chapter 1 for comments.)

If, for example, $D_1 = D_2 = D$; $U_1 = U_2 = U$; $\rho_1 = \rho_2 = \rho$; and $A_1 = A_2 = A_q = A$, then it follows from Equations (2.26) and (2.27) that

$$\frac{\partial v_1}{\partial z} - \frac{\partial v_2}{\partial z} = \frac{q}{D} \quad \text{at} \quad z = 0 \tag{2.28}$$

The Laplace transforms of Equation (2.24) and conditions (2.25a), (2.25b), (2.27), and (2.28) give

$$\begin{aligned}\bar{v}_1 = {} & q_{\text{pl}}/\{2Ds[(U^2/4D^2) + (s/D)]^{1/2}\} \\ & \times \exp\{(U/2D) + [(U^2/4D^2) + (s/D)]^{1/2}\}z\end{aligned} \tag{2.29}$$

and

$$\begin{aligned}\bar{v}_2 = {} & q_{\text{pl}}/\{2Ds[(U^2/4D^2) + (s/D)]^{1/2}\} \\ & \times \exp\{(U/2D) - [(U^2/4D^2) + (s/D)]^{1/2}\}z\end{aligned} \tag{2.30}$$

From the table of transforms these can be expressed as integrals (see Appendix 5), and these turn out to be the same as Equation (2.21). Their evaluation then follows the same lines. It is to be noticed that in contrast to the instantaneous source (to be discussed next), the concentration in the case of a continuous source does not tend to infinity for $t \to 0$, $z \to 0$.

2.3.5 An Instantaneous Plane Source in a Moving Medium

Consideration of the *instantaneous* plane source in a moving medium has been deferred, and so out of its logical order, because it has taken on an importance such that the remainder of this chapter as well as subsequent chapters are devoted to it.

(a) *The Use of the Concept of Kelvin's Instantaneous Source* [2]

An instantaneous plane source of strength Q_{pl} in an isotropic, stationary medium gives rise to a concentration distribution

$$v(Z, t) = [Q_{\text{pl}}/2(\pi Dt)^{1/2}] \exp(-Z^2/4Dt) \tag{2.31}$$

(where Z is the distance measured from the source, which is therefore taken to be the origin) in that it satisfies Fourier's differential equation

$$\frac{\partial^2 v}{\partial Z^2} - \frac{1}{D}\frac{\partial v}{\partial t} = 0 \tag{2.32}$$

and the conditions

$$v \to 0 \quad \text{for} \quad t \to 0 \quad \text{for} \quad Z \neq 0 \tag{2.33}$$

$$v \to \infty \quad \text{for} \quad t \to 0 \quad \text{for} \quad Z = 0 \tag{2.34}$$

$$v \quad \text{is finite for} \quad Z \neq 0 \tag{2.35}$$

$$\int_{-\infty}^{\infty} v\, dZ = Q_{\text{pl}} \tag{2.36}$$

If, now, this line of reasoning is applied to the case where the medium moves in the positive direction with velocity U, then the concentration at position z—where z is measured from an origin that is the position in space where the source irrupted—can be obtained at any time t by writing $(z - Ut)$ for Z in Equation (2.31); i.e., Z now represents distances measured not from the fixed origin, but from one that moves with the stream and which was at the point where the source irrupted at an epoch that is taken to be the origin of time. This is illustrated in Figure 2.2.

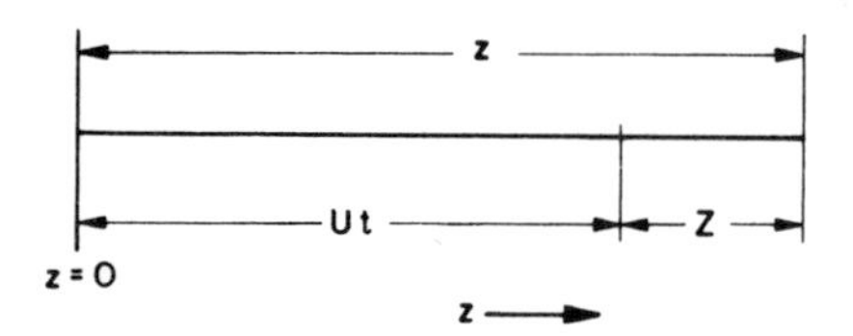

Fig. 2.2. Relation among Z, z, and t.

The concentration at any point that is at a distance z from the fixed origin and at time t will therefore be

$$v(z, t) = [Q_{\mathrm{pl}}/2(\pi Dt)^{1/2}] \exp -[z - Ut)^2/4Dt] \qquad (2.37)$$

(b) The Direct Solution of the Differential Equation

The differential equation applicable to this case is

$$\frac{\partial^2 v}{\partial z^2} - \frac{U}{D}\frac{\partial v}{\partial z} - \frac{1}{D}\frac{\partial v}{\partial t} = 0 \qquad (1.5\mathrm{b})\ \mathrm{R}$$

with the conditions (2.33)–(2.36) but with Z replaced by z. As shown in Appendix 2, the substitution $Z = z - Ut$ converts the differential equation (1.5b) into one of the form of Equation (2.32), and the conditions to forms identical to Equations (2.33)–(2.36), but the variable Z now signifying $z - Ut$. It will be found that the solution of Equation (2.32) with conditions (2.33)–(2.36) is the same as Equation (2.37) after $z - Ut$ has been resubstituted for Z. (The Laplace transform of the solution will be found to be

$$\bar{v}(s, Z) = [(Q_{\mathrm{pl}}/2D)(D/s)^{1/2}] \exp[-(s/D)^{1/2}Z] \qquad (2.38)$$

Appendix 6 contains a note on this.)

REFERENCES

[1] O. Levenspiel, and K. B. Bischoff, "Advances in Chemical Engineering" (T. B. Drew, J. W. Hoopes, Jr., and T. Vermeulen, eds.), Vol. 4. Academic Press, New York, 1963.

[2] H. S. Carslaw, and J. C. Jaeger, "Conduction of Heat in Solids," 2nd. ed. Oxford Univ. Press, London and New York, 1959.
[3] H. A. Wilson, On Convection of Heat. *Proc. Cambridge Phil. Soc.* **12**, 406 (1904).
[4] G. A. Gibson, "An Elementary Treatise on the Calculus." Macmillan, New York, 1946.

Chapter 3

The Ideal Impulse in One-Dimensional Flow

The last chapter ended with the derivation of the concentration as a function of time and of distance from a fixed origin that arises from an ideal, plane impulse in one-dimensional flow and was given by Equations (2.37) and (2.31).

3.1 THE VARIATION OF CONCENTRATION WITH DISTANCE FROM A MOVING ORIGIN

The concentration function is symmetric in space (about $z = Ut$, i.e., the point in the flow that was at the origin when the source irrupted) at any given instant of time, and is the "normal" or "Gaussian" error curve [1].

The integrated value from $z \longrightarrow -\infty$ to $z \longrightarrow +\infty$ has already been stated in Equation (2.36) as Q_{pl}. Hence, if the theoretical curve is to be fitted to an experimental one by adjusting the value of some parameter so that the curves overlap at all points, then the magnitude of this parameter can be found. This was done for example, by Taylor [2], who in this way found a value of D (in order to test his theory by which the value of D could be computed in slow flow in a capillary tube). In practice it is usually possible to get only portions of the experimental and theoretical bell-shaped curves to overlie one another; it is therefore worth considering whether some parts of the curve are likely to give more accurate results than others. For example, if the value of D is the major uncertainty, then differentiation of Equation (2.31) with respect to D gives

$$\Delta v/v \approx [(Z^2/4Dt) - \tfrac{1}{2}]\, \Delta D/D \tag{3.1}$$

that is to say, in the curve that shows concentration varying with Z, the fractional change of the ordinate that varying D brings about is equal to some factor times the fractional change in the value of D that causes it. The larger the value of the factor, the more sensitive will be the change in the shape of the curve to a given fractional change in the value of D; this, however, is not the only criterion, as will be seen when it is considered what happens at different values of Z. Thus, when $Z = 0$, the factor is equal to $-\frac{1}{2}$, and as Z increases (in absolute value) the factor decreases (in absolute value), becoming zero at $Z = (2tD)^{1/2}$ and then becomes positive and larger as Z increases, tending to infinity as Z so tends. Nevertheless, for these large values of Z the *absolute* value of Δv tends to

$$\Delta v \longrightarrow (Z^2/4tD)[\exp(-Z^2/4Dt)][Q/2(\pi Dt)^{1/2}]\, \Delta D/D \tag{3.2}$$

and this tends to zero as Z tends to infinity, as is shown in books of mathematical analysis (e.g., Hardy [3]) and it is this absolute value which has to

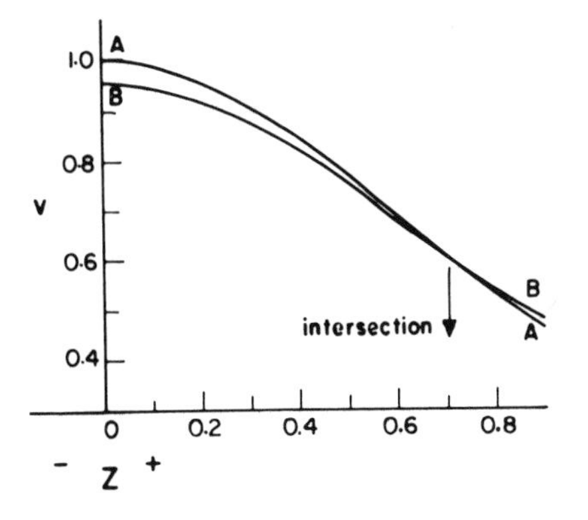

Fig. 3.1. Concentration from instantaneous plane source [Equation (2.31)]. Curve A: $Q_{pl}/\pi^{1/2} = (4Dt)^{1/2} = 1$. Curve B: Value of D increased by 10%. Curves intersect at $Z = 1/\sqrt{2}$.

be gauged, often by eye. Furthermore, the experimental curve itself is often inaccurate at these large values of Z, because of the difficulties of measurement. To illustrate the argument, Figure 3.1 shows two curves plotted from Equation (2.31): curve A is for $4Dt = 1$ and curve B is for $4Dt = 1.1$. At $Z = 0$ both the relative and absolute changes in Δv are large. It is best therefore to deal with the region from $Z = 0$ to $Z = \pm 2(2tD)^{1/2}$ if possible.

3.2 THE VARIATION OF CONCENTRATION WITH TIME

A more widely used—because it is easier—practice is to measure the variation of concentration with time at a station that is at some fixed distance z from the origin of the instantaneous source. The distribution so found is an asymmetric one, which exhibits limiting shapes as the dispersion coefficient D is varied, and which are shown in Figure 3.2. The total area under the curve is

$$\int_{t'}^{\infty} v\,dt = \int_{0}^{\infty} v\,dt$$

where t' is absolute time and the source irrupted at $t' = t$ on this scale; the magnitude of this area can be arrived at by remembering that in the small time dt the fluid has moved a distance $U\,dt$. Hence, in all time from zero to

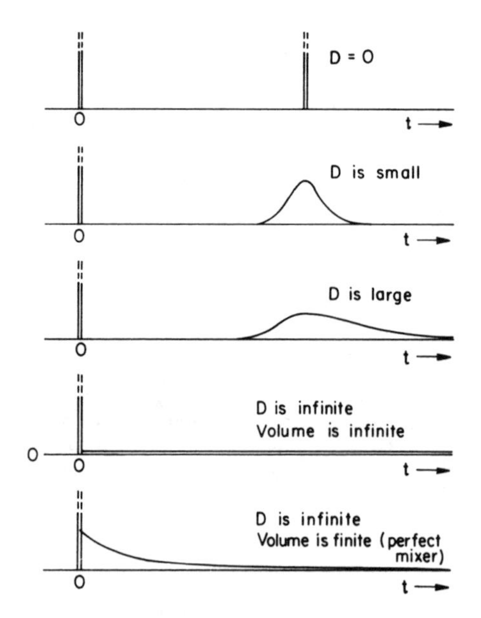

Fig. 3.2. Effect on Dirac impulse of various flow situations.

infinity an infinitely long column of fluid will have flowed. Since

$$\int_{-\infty}^{\infty} v\, dz = Q_{\text{pl}}$$

by Equation (2.36), it follows that

$$\int_{t'}^{\infty} v\, dt' = \int_{0}^{\infty} v\, dt = \int_{-\infty}^{\infty} v(dt/dz)\, dz = Q_{\text{pl}}/U \tag{3.3}$$

At the limiting value of $D = 0$ physical reasoning indicates that the plane source should suffer no distortion and should therefore reappear at the measuring station, still in the form of an instantaneous source, at a time $t = z/U$ later. That this is so is shown by substituting $D \longrightarrow 0$ into Equation (2.31); v approaches zero for all Z except $Z = 0$; i.e., except at $z = Ut$.

On the other hand, substitution of $D \longrightarrow \infty$ into Equation (2.31) leads to $v \longrightarrow 0$ for all Z. Yet this seems to be an example of a *perfect mixer*, in which the concentration rises instantaneously to some value, then to decline exponentially. (See, for example, Levenspiel [4].) The reason it does not do the same thing in this case is, of course, that the perfect mixer, to behave as described, must be a system of limited volume with precise boundaries and clearly defined boundary conditions. The present system has been defined to be of infinite volume; the analytic result is therefore seen to be correct.

When D has a finite value the results are of greater interest. Figures 3.3 and 3.4 show that the magnitude of D influences the shape of the curve appreciably, even when the range of values is finite; the maxima lie on the dotted line in Figure 3.4. Since the *magnitude* of D can only have a meaning when it is compared with something, it is best to rearrange Equation (2.37) into the form

$$\begin{aligned} vz/Q = {} & \{1/[2(\pi)^{1/2}(D/Uz)^{1/2}(Ut/z)^{1/2}]\} \\ & \times \exp\{-[1 - (Ut/z)]^2/4(D/Uz)(Ut/z)\} \end{aligned} \tag{3.4}$$

It will be seen that each of the groups in parentheses is now dimensionless, and for convenience each will be denoted by a symbol and a name, viz.,

$$D/Uz = 1/\text{Pe} \tag{3.5}$$

the inverse of the Peclet number;

$$vz/Q = \nu \tag{3.6}$$

the dimensionless concentration;

$$Ut/z = \theta \tag{3.7}$$

the dimensionless time ($= 1/\chi$); [also

$$z/Ut = \chi \tag{3.8}$$

the dimensionless distance ($= 1/\theta$)].

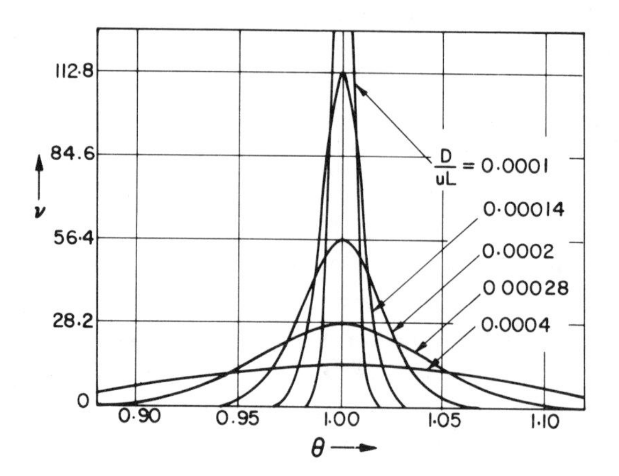

Fig. 3.3. Effect of value of Peclet number on Dirac impulse (large values of Peclet number). (With acknowledgments to Levenspiel and Smith [5].)

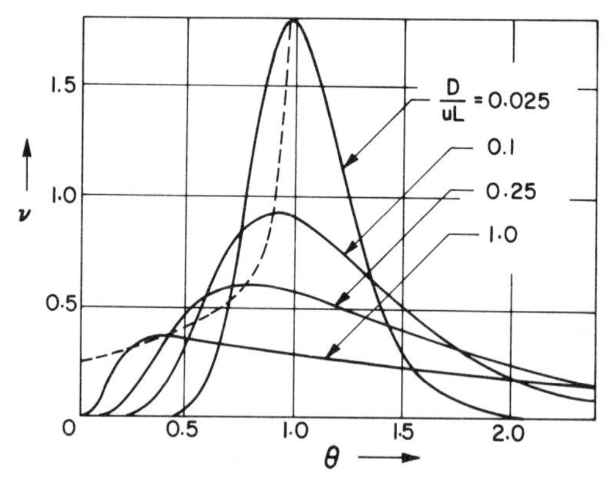

Fig. 3.4. Effect of value of Peclet number on Dirac impulse (small values of Peclet number). (With acknowledgments to Levenspiel and Smith [5].)

The dimensionless concentration ν is the ratio of the actual concentration at z to what it would be if the amount in the instantaneous source were distributed uniformly in a volume of unit cross-sectional area and of length z. The dimensionless time θ is the ratio of the time t to the time needed for the volume contained between $z = 0$ and $z = z$ to clear the system; this is sometimes called the time of one displacement. The dimensionless distance χ, the reciprocal of the dimensionless time, is the ratio of the actual distance z to the distance the fluid has moved in time t.

In terms of these newly defined quantities Equation (3.4) can be written as

$$\nu = (\mathrm{Pe}^{1/2}/2\pi^{1/2}\theta^{1/2}) \exp -[\mathrm{Pe}(1 - \theta)^2/4\theta] \qquad (3.9)$$

or

$$\nu = (\mathrm{Pe}^{1/2}/2\pi^{1/2}\theta^{1/2}) \exp -[\mathrm{Pe}(1 - \chi)^2/4\chi] \tag{3.10}$$

Equation (3.9) is perhaps the more convenient one to consider: it expresses the relation between ν and θ in terms of Pe, and this relationship is shown graphically in Figures 3.3 and 3.4, which show how the shape of the curve depends on the magnitude of Pe. As Pe increases, the curve becomes more symmetric, with a pronounced peak. The graphs are from Levenspiel and Smith [5], who point out that when Pe is very large (for example, $\mathrm{Pe} > 10^2$), ν is always small except when $1 - \theta$ is small, as can be seen by examining the exponent in Equation (3.9) [see the remark following Equation (3.2)]. Thus, $\theta = 1 + \Delta\theta \approx 1$, where $\Delta\theta$ is small; so, when Pe is large

$$\nu \approx (\mathrm{Pe}^{1/2}/2\pi^{1/2}) \exp[-\mathrm{Pe}(\Delta\theta)^2/4] \tag{3.11}$$

The right-hand side of this last equation is the normal probability curve (Appendix 4), showing that ν is symmetric about $\theta = 1$. The physical meaning of this is that the shape of the concentration versus distance curve changes but little as the wave passes the observation station, as Figure 3.3 (which refers to cases where $\mathrm{Pe} > 10^3$) shows [5].

In contrast, for values of Pe smaller than 10^3, and particularly when Pe is smaller than 10^2, the concentration versus distance curve is broadening rapidly as time goes on. That is to say, the fixed observer is "seeing" not merely a pulse passing the station, but a pulse that is rapidly changing in shape as it flows. The pulse then first approaches as a steeply rising change in concentration; it departs as a more gently declining one whose "tail" is stretching out because of dispersive effects, and Figure 3.4 illustrates this case.

3.3 THE USE OF CONCENTRATION DISTRIBUTIONS TO MEASURE MEAN VELOCITIES

As stated, the curve showing how the concentration varies with time when an instantaneous source irrupts upstream of the observation point will have a shape dependent on the value of D. If there is no dispersion, i.e., if $\mathrm{Pe} \longrightarrow \infty$, then the observer at a distance L downstream of the source "sees" it at a time t later, and so $U = L/t$. If Pe has a finite value, however, then the pulse passes the observer as a more or less long drawn-out one: there is really no end to it. The questions arise: Can the mean velocity of flow now be determined, and what time would then be used to compute it? In Chapter 2 a method was mentioned of finding U by a continuous injection of matter into the stream; the use of an instantaneous source may provide an alternative method. If so, only a relatively small amount of tracer would be needed.

3.3.1 The Time at the Peak of the Curve

The easiest time to measure is that elapsing between the source irrupting and the peak of the subsequent wave passing the observer: this is denoted by t_{peak}. Because it is such an easy measurement to make, it is often used to determine the flow velocity [6], but it may be suspected that the reasons for its use are this ease of determination—and intuition.

Figure 3.4 shows that, in general, the time t_{peak} is not equal to L/U (i.e., $\theta_{peak} \neq 1$). It is approximately so if Pe is large; Levenspiel and Smith [5] have examined this and they point out that differentiation of Equation (3.9) gives

$$\theta_{peak} = (\mathrm{Pe}^{-2} + 1)^{1/2} - \mathrm{Pe}^{-1} \tag{3.12}$$

So θ_{peak} lies between 0 and 1: its value as a function of Pe is shown in Figure 3.6. It can be seen from this that when Pe is greater than about 10 then $\theta_{peak} \doteqdot 1$, but that when Pe is small, say $\mathrm{Pe} < 0.6$, then θ_{peak} is a function of Pe; namely $\theta_{peak} \approx \mathrm{Pe}/2$, as is shown in Section 3.4.1. It therefore has to be decided whether Pe is large or small before the merits of using θ_{peak} as a means of measuring the velocity can be judged.

The values of ν at the peak (viz., ν_{peak}) can be obtained from Equations (3.12) and (3.9) as

$$\nu_{peak} = [(1 + \mathrm{Pe}^2)^{1/2} + 1]^{1/2}/2\pi^{1/2} \times \exp\{-[(1 + \mathrm{Pe}^2)^{1/2} - \mathrm{Pe}]/2\} \tag{3.13}$$

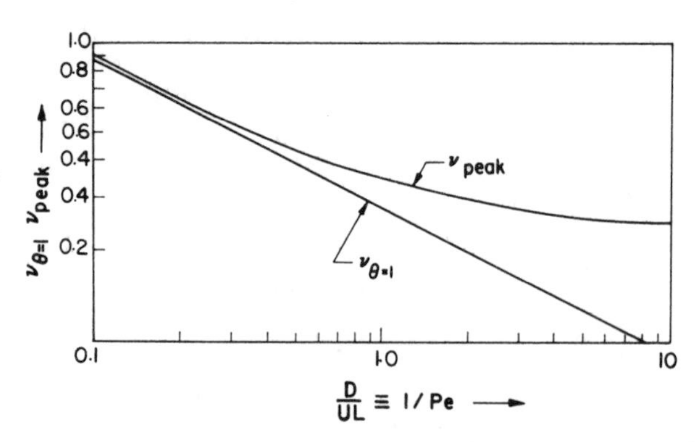

Fig. 3.5. Effect of value of Peclet number on concentrations at peak and at $\theta = 1$ (degenerate impulse). (With acknowledgements to Levenspiel and Smith [5].)

Figure 3.5 shows how these values of ν_{peak} vary as Pe varies. It also shows the values of $\nu_{\theta=1}$ that are obtained by putting $\theta = 1$ into Equation (3.9), getting

$$[\nu(\text{Pe})]_{\theta=1} = \text{Pe}^{1/2}/2\pi^{1/2} \tag{3.14}$$

If the value of Pe is known (for it is possible to determine values of Pe without knowing U or L, as will be shown), then another way of finding the velocity is to use Figure 3.5 to find the ratio of ν_{peak} to $\nu_{\theta=1}$ $(= v_{\text{peak}}\colon v_{\theta=1})$; so the point on the curve of v versus t (in the direction of t increasing) where v has the value $v_{\theta=1}$ (found from Figure 3.5) will be at $t_{\theta=1}$, so $U = L/t_{\theta=1}$. The time $t_{\theta=1}$ is the same as $t_{1/2}$, to be discussed in the next section.

3.3.2 The Time That Halves the Area under the Trace

Having dealt with the method of finding U that requires the least information (viz., a part of the concentration–time record that covers the peak), we now turn to consider values of a time that has to be found by a lengthier process.

The first—considered in this section—is that time that divides the total area under the v versus t curve into exact halves. (The second method, which will be dealt with in the next section, is the value of the "average" time.)

Let the time that halves the area under the complete concentration–time trace (measured at a distance L from the source) be denoted by $t_{1/2}$. Since it has already been shown that the curve of concentration versus Z (where $Z = z - Ut$) is symmetric about $Z = 0$, then we can write

$$\begin{aligned}\tfrac{1}{2} &= \int_0^\infty v\,dZ \Big/ \int_{-\infty}^\infty v\,dZ \\ &= \int_{t_{1/2}=L/U}^{-\infty} v(\partial Z/\partial t)_{z=L}\,dt \Big/ \int_\infty^{-\infty} v(\partial Z/\partial t)_{z=L}\,dt \\ &= U\int_{t_{1/2}=L/U}^\infty v\,dt \Big/ U\int_0^\infty v\,dt\end{aligned}$$

since

$$\int_{-\infty}^0 v\,dt = 0 \qquad \text{and} \qquad Z = z - Ut$$

hence $(\partial Z/\partial t)_z = -U$; but $t_{1/2}$ has been chosen to make the area under the concentration versus time curve one-half of the area under the complete curve; so from the value of the lower limit of integration in the above it follows that $U = L/t_{1/2}$. Since now $t_{1/2}/UL = \theta = 1$, it follows that $t_{1/2}$

is the same time as that mentioned at the end of the Section 3.3.1. It can be found by plotting the cumulative area under the curve as a function of t.

3.3.3 The Mean Residence Time

The mean residence time $\bar{t}$ is defined to be

$$\bar{t} = \int_0^\infty vt\,dt \Big/ \int_0^\infty v\,dt$$

where v is, in this case, anything proportional to the concentration caused by an instantaneous source in an infinite medium, while t is in units of time. There might be a temptation to think that this could give the velocity of flow as $U = L/\bar{t}$. This is not so; it is worth finding out why, whether it can be related to anything else, and whether it has a use in the present context. An equation for $\bar{t}$ can be found by substituting Equation (2.37) into the numerator and using Equation (3.3) for the denominator:

$$\bar{t} = [Q_{pl}/2(\pi D)^{1/2}] \int_0^\infty t^{1/2} \exp[-(z - Ut)^2/4Dt]\,dt \Big/ (Q_{pl}/U) \qquad (3.15)$$

The evaluation of the integral is shown in Appendix 3 and leads to the mean residence time $\bar{t}$ at a point at a distance L from the source in a medium flowing at velocity U being given by

$$\bar{t} = (L/U)(1 + 2\mathrm{Pe}^{-1}) \qquad (3.16)$$

or

$$\bar{\theta} = (1 + 2\mathrm{Pe}^{-1}) \qquad (3.17)$$

where $\bar{\theta}$ is the dimensionless mean time in reduced units, and so $U = L/\bar{t}$ only if Pe is large, the same condition that governed the use of t_{peak}.

3.3.4 The Determination of the Velocity of Flow by an Instantaneous Source: Conclusions

The magnitude of the "times" dealt with in Section 3.3 are shown in Figure 3.6 as functions of Pe, and Figure 3.7 shows diagrammatically these quantities on a typical curve of concentration versus θ. The most easily measured, θ_{peak}, is only accurate if Pe is large. The other two require complete records of the concentration versus time: of these, $\theta_{1/2}$ gives a theoretically correct answer, while $\bar{\theta}$ gives an answer that is correct only when Pe is large and in addition is likely to be the more inaccurate because in the "tail," errors of measurements will be magnified by their being multiplied by these large values of t. An example is given in Chapter 9.

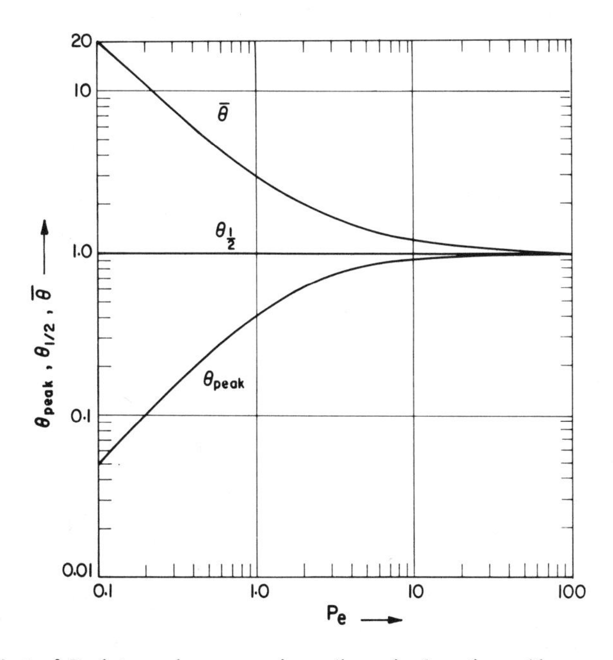

Fig. 3.6. Effect of Peclet number on various dimensionless times (degenerate impulse).

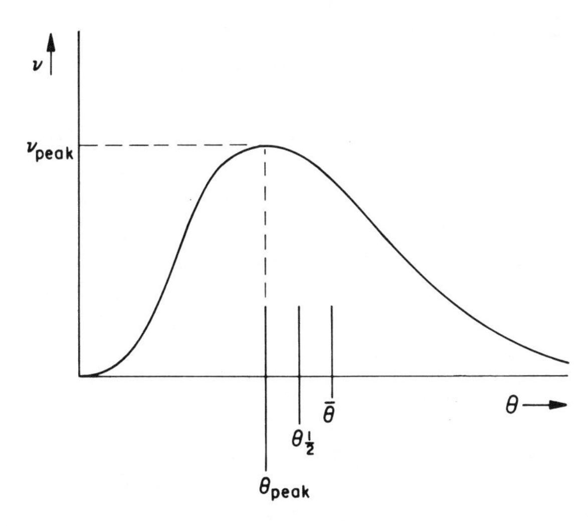

Fig. 3.7. Relation of various dimensionless times to one another (degenerate impulse). (Diagrammatic only.)

3.4 THE USE OF CONCENTRATION DISTRIBUTIONS TO MEASURE THE PECLET NUMBER

3.4.1 Small Values of Pe: The Use of θ_{peak} and of $\bar{\theta}$

Two relations have already been derived between the Peclet number and a time that can be computed from a concentration–time distribution (determined at a distance L downstream of an instantaneous source); these are

$$\theta_{peak} = (1 + \text{Pe}^{-2})^{1/2} - \text{Pe}^{-1} \qquad (3.12)\ \text{R}$$

and

$$\bar{\theta} = (1 + 2\text{Pe}^{-1}) \qquad (3.17)\ \text{R}$$

So the value of Pe can be computed if either θ_{peak} or $\bar{\theta}$ is calculated, although if Pe is large, the determination is inaccurate because of the small differences between large quantities that are invoked. Again, because of the long tail of the distribution, greater accuracy is often obtained by using θ_{peak} rather than $\bar{\theta}$, as pointed out in Subsection 3.3.4. By a slight rearrangement of Equation (3.12), there results

$$\theta_{peak} = \{[1 + (\text{Pe}^2/2) - (\text{Pe}^4/8) + \cdots] - 1\}/\text{Pe}$$

Hence, when Pe is small

$$\theta_{peak} = Ut_{peak}/L \approx \text{Pe}/2 \qquad (3.18)$$

An appreciation of the magnitude of the error caused by the approximation in Equation (3.18) can be gained by examination of Figure 3.6.

3.4.2 Large Values of Pe: The Use of the Normal Error Curve

If Pe is large, then the method in Section 3.4.1 is inaccurate; on the other hand, the concentration distribution with time is now almost of Gaussian "normal" shape, and this, due to its simple analytic definition, has many useful attributes. In particular, if anything proportional to concentration is plotted or recorded against θ as the independent variable (for which both L and U would have to be known), then at an abscissa distant θ_1 from the axis of symmetry, such that $v = 0.3679v_{max}$, the Peclet number is given by $\text{Pe} = 4/\theta_1^2$; this derives from the facts that $\theta_1 = 2^{1/2}\sigma$ (by Appendix 4) and $\text{Pe} = 2/\sigma^2$ when Pe is large, by Equation (3.27). Figure 3.8 shows the method.

Equally, a reduced concentration would serve. Thus, Equation (3.13) or (3.14) could be used, or, more conveniently, the value of Pe could be read off from Figure 3.5, if either ν_{peak} or $\nu_{\theta=1}$ is computed.

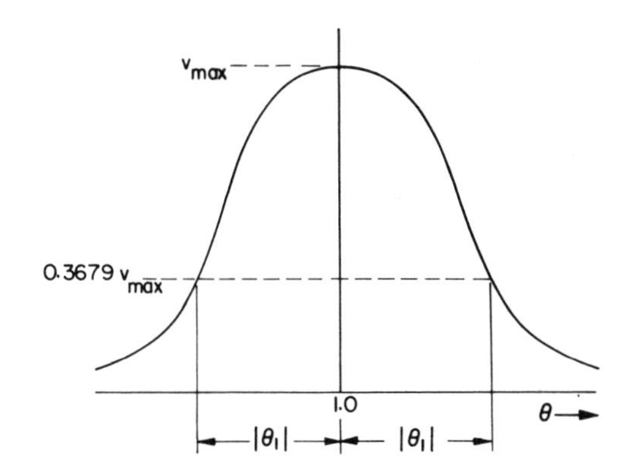

Fig. 3.8. Method used to find the Peclet number ($= 4/\theta_1{}^2$) if pulse has shape of a normal error curve.

3.4.3 The Use of the Variance of the Concentration Distribution

In Appendix 7 it is shown that the statistical parameter, the variance (denoted by σ^2), can be calculated from a concentration–time distribution, for such a curve could be considered to be a measure of the distribution of probabilities that the small parcel of fluid which has been introduced into the system at $z = 0$ and $t = 0$ would be at $z = L$ at $t = t$. Another parameter is needed, namely $\bar{t}$, already defined as $\bar{t} = \int_0^\infty ct\, dt$. This enables times to be measured from any origin in the calculation of the variance, which is defined to be

$$\sigma^2 = \int_0^\infty c(t - \bar{t})^2\, dt \tag{3.19}$$

provided that c (a quantity proportional to concentration. i.e., $c = \tilde{A}v$, where $\tilde{A}$ is a constant) is such that $\int_0^\infty c\, dt = 1$; i.e., there is certainty that, given enough time, every minute parcel of fluid that was at $z = 0$ at $t = 0$ will arrive at $z = L$. So, if actual concentrations are to be used, then

$$\sigma^2 = \tilde{A} \int_0^\infty v(t - \bar{t})^2\, dt \qquad \text{and} \qquad \tilde{A} \int_0^\infty v\, dt = 1$$

or

$$\tilde{A} = 1 \Big/ \int_0^\infty v\, dt \tag{3.20}$$

So,

$$\sigma^2 = \int_0^\infty v(t - \bar{t})^2\, dt \Big/ \int_0^\infty v\, dt \tag{3.21}$$

It is also shown in Appendix 7 that Equation (3.19) can be rewritten as

$$\sigma^2 = \int_0^\infty ct^2\, dt - \left(\int_0^\infty ct\, dt\right)^2 \tag{3.22}$$

If Equation (3.20) is used in conjunction with Equation (3.22), then

$$\sigma^2 = \left(\int_0^\infty vt^2\, dt \Big/ \int_0^\infty v\, dt\right) - \left(\int_0^\infty vt\, dt \Big/ \int_0^\infty v\, dt\right)^2 \qquad (3.23)$$

The second term bracket is the mean time $\bar{t}$, which has been shown in Section 3.3.3 to be given by

$$\bar{t} = (L/U)(1 + 2\text{Pe}^{-1}) \qquad (3.16)\ \text{R}$$

As for the first term this can be found by substituting for v from Equation (2.37); the subsequent integration uses the result shown in Appendix 3, while Equation (3.3) provides the denominator. The result is

$$\sigma_t^2 = (L/U)^2(8\text{Pe}^{-2} + 2\text{Pe}^{-1}) \qquad (3.24)$$

or

$$\sigma_\theta^2 = (8\text{Pe}^{-2} + 2\text{Pe}^{-1}) \qquad (3.25)$$

The dimensions of σ^2 are: (dimensions of the observed quantity)^2; hence. they are (time)^2 in the first and are zero in the second of these last two equations.

Equation (3.25) can be rearranged to

$$\text{Pe} = [(1 + 8\sigma_\theta^2)^{1/2} + 1]/\sigma_\theta^2$$

3.4.4 Large Values of Pe

If Pe is so large that in Equation (3.25) $8\text{Pe}^{-2} \ll 2\text{Pe}^{-1}$, i.e., if $\text{Pe} \gg 4$, say,

$$\text{Pe} \geq 40 \qquad (3.26)$$

then

$$\text{Pe} = 2/\sigma_\theta^2 \qquad (3.27)$$

a result already used in Section 3.4.2.

[Note: A numerical value of the variance can always be found, whatever the shape of the distribution; i.e., whether it obeys Equation (2.37) or not. The significance of this more general case will be considered in Chapters 4 and 5.]

3.4.5 Some Remarks on Errors in the Determinations

If the errors in measuring the variables are relatively small, then the values of the fractional errors in θ and v can be deduced from their definitions. Thus, for θ: since $\theta = \theta(U, t, z)$, it follows that

$$d\theta = \frac{\partial\theta}{\partial U} dU + \frac{\partial\theta}{\partial t} dt + \frac{\partial\theta}{\partial z} dz$$

and since $\theta = Ut/z$, and all these quantities are measured separately and thus invite errors in each, it follows that the fractional error in θ is

$$\frac{d\theta}{\theta} \doteqdot \frac{dU}{U} + \frac{dt}{t} - \frac{dz}{z} \tag{3.28}$$

The last are the fractional errors in U, t, and z, their signs being taken into account.

Similarly, for ν: since $\nu = vz/Q$, it follows that

$$\frac{d\nu}{\nu} \doteqdot \frac{dv}{v} + \frac{dz}{z} - \frac{dQ}{Q} \tag{3.29}$$

These are errors due to the "unexplainable" or chance errors in measurement. It will be seen that relatively modest errors could mount up to a sizable one in θ, and (particularly if $v \equiv$ heat, obtained by measuring temperature, density, and specific heat) in ν. Further, the relative accuracy of Pe will depend on these relative accuracies in ways that can be found by using Equation (3.12). In this way one gets

$$d(\text{Pe})/\text{Pe} = (1 + \text{Pe}^2)^{1/2}(d\theta/\theta)_{\text{peak}} = \tilde{P}(d\theta/\theta)_{\text{peak}} \tag{3.30}$$

and $\tilde{P}$ is shown in Figure 3.9 as $\tilde{P} = (1 + \text{Pe}^2)^{1/2}$. Hence, if Pe is large, then the fractional error can be many times that in θ_{peak} (as can be deduced from the shape of the curve for large values of Pe in Figure 3.6).

Again, the relation between the relative errors in $\nu_\theta = 1$ and Pe may be obtained from Equation (3.14): it is simply

$$d(\text{Pe})/\text{Pe} = 2\, d\nu_{\theta=1}/\nu_{\theta=1} \tag{3.31}$$

Finally, the rather more complicated relation between the relative errors in ν_{peak} and Pe may be obtained from Equation (3.13). It is

$$d(\text{Pe})/\text{Pe} = (d\nu_{\text{peak}}/\nu_{\text{peak}})f \tag{3.32}$$

where

$$f = \{[(1 - 2\tilde{P})/(\tilde{P})]/2\text{Pe}\} + [1/\tilde{P}(\tilde{P} + 1)] \tag{3.33}$$

where $\tilde{P} = (1 + \text{Pe}^2)^{1/2}$. Figure 3.10 shows how f varies with Pe.

When the value of Pe is being calculated in the other way that has been described, viz., from the variance of the concentration distribution, the relative error can be obtained immediately from Equation (3.25) as

$$d(\text{Pe})/\text{Pe} = -[(4\text{Pe}^{-1} + 1)/(8\text{Pe}^{-1} + 1)]\,(d(\sigma_\theta^2)/\sigma_\theta^2) \tag{3.34}$$

As regards the errors in determining σ^2, the subject must be referred to practical texts on statistics; in brief, it has already been pointed out that a long "tail" of concentration as t increases causes any errors in determining

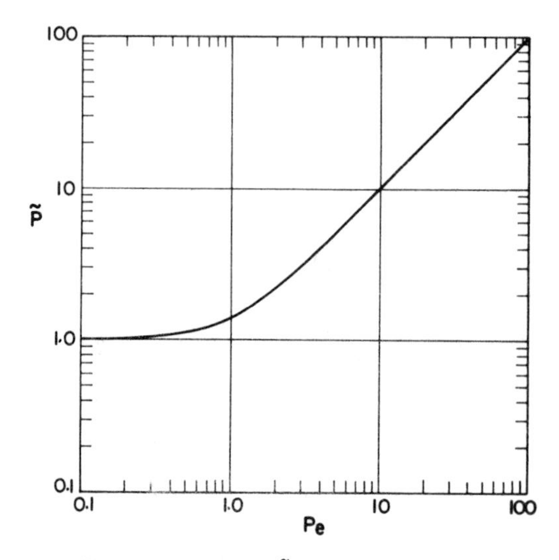

Fig. 3.9. Parameter $\tilde{P}$ as a function of Pe.

v to be magnified. Hence, it follows that errors in σ^2 will be reduced if the distribution is humplike. This calls for a small value of Pe. Yet, for small values of Pe the factor in the bracket becomes larger: however, it is not very sensitive to values of Pe, since it ranges only between $\frac{1}{2}$ and its maximum value of 1.

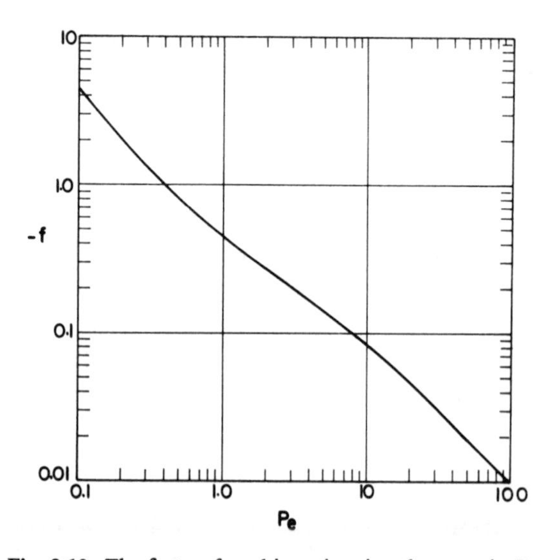

Fig. 3.10. The factor f used in estimating the error in Pe.

3.5 EXAMPLES OF THE USE OF AN INSTANTANEOUS SOURCE TO DETERMINE Pe (BASED ON EXAMPLES BY LEVENSPIEL AND SMITH [5])

(1) *Using the Value of* $\nu_{\theta=1}$

Water flows at 1.17 ft sec^{-1} in a stream tube of cross-sectional area denoted by A: its value need not be known. An impulse of total magnitude Q (equivalent to $Q_{\text{pl}} = Q/A$) was injected: the material used was a tracer of potassium permanganate, the amount being reported as being enough to fill a 1 in. length of tube, and the concentration, determined colorimetrically, was denoted by 1 on an arbitrary scale. This was equivalent to $1 \times v_Q$ in some arbitrary units, where v_Q is the proportionality constant.* At a point 9 ft downstream the concentration of the solution was determined, on the same arbitrary scale as before, at a time $9/1.17 = 7.7$ sec later; thus, $\theta = Ut/L = 1$. The value of the measured concentration was $0.00555v_Q$. Since $Q = Av_Q(1/12)$ (arbitrary units $\times$ ft^3), then $\nu_{\theta=1} = (0.00555v_Q \times 9)/[(1/12)Av_Q/A] = 0.60$. From Figure 3.5 it is found that Pe $= 4.61$ when $v_{\theta=1} = 0.6$. Hence, $D = (1.17 \times 9)/4.6 = 2.28$ ft^2 sec^{-1}.

Quantities required	Quantities not required
Concentration (in arbitrary units)	Actual concentration
The magnitude of the impulse as a volume of fluid (per unit area of cross section) of a given concentration (arbitrary units)	Actual cross-sectional area
L	
U	

(2) *Using the Value of* ν_{peak}

Some values of concentration were taken near the peak of the distribution: it was found that $\nu_{\text{peak}} = 65$ (in arbitrary units). [This occurred 6.2 sec after the instant that the impulse was injected: this time is not required for the present calculation, but it should be compared with the time in (1).] Now, $\nu_{\text{peak}} = v_{\text{peak}}L/Q_{\text{pl}}$, and so the value of Q_{pl} is needed. This may be

*Levenspiel and Smith [5] gave concentration v_Q as being 1 % w/w, but it is not necessary to know this.

found in one of two ways: (a) by injecting a known amount of material (or heat) per unit of cross-sectional area, in which case v needs to be known in the same units as were used to compute Q_{pl} (see Example 1); or (b) by computing the area under the curve of concentration versus time. For, $U \int_0^\infty v\, dt = Q_{pl} =$ the total amount (per unit area) passing through the system.

Calculations

METHOD (a)

From Levenspiel and Smith's results the values of concentration on the arbitrary scale have to be converted to the scale used in determining Q_{pl} in the following manner (their results are given in Table 3.1): At 7.7 sec, the concentration was reported to be 0.00555% w/w, and the value on the arbitrary scale estimated from their data was 59: hence, v_{peak}, at about 65 arbitrary units, would be equivalent to 0.0060% w/w. Hence, $\nu_{peak} = (0.0060 \times 9)/(1 \times 1/12) = 0.65$, and from Figure 3.5 it is seen that Pe $= 4.61$, as before.

TABLE 3.1

CONCENTRATION OF POTASSIUM PERMANGANATE VERSUS TIME AT A FIXED OBSERVATION POINT [5]

Time (sec)	Concentration of KM_nO_4[a]	Time (sec)	Concentration of KM_nO_4[a]	Time (sec)	Concentration of KM_nO_4[a]
0	0	16	22	30	2
2	11	18	16	32	2
4	53	20	11	34	2
6	64	22	4	36	1
8	58	24	7	38	1
10	48	26	5	40	1
12	39	28	4	42	1
14	29				

[a]Arbitrary units.

METHOD (b)

In order to find the area under the curve, use is made of the information in Example 2 of Levenspiel and Smith's paper. Here, from the measured values, it can be computed that $\sum v = 386$ (where v is the symbol for concentration in arbitrary units). Since these were at 2-sec intervals, and since these are fairly small when judged with reference to the curvature of the distribution, a good approximation to the area was 2×386 (arbitrary units of concentration $\times$ seconds). Hence, the total amount of solute was proportional to $1.17 \times 2 \times 386$ and so $\nu_{peak} = (65 \times 9)/(1.17 \times 2 \times 386) = 0.65$, as for (a). (It will be seen that the units of concentration are the same in both numerator and denominator.)

Required	Not required
Example 2(*a*)	
L	t_{peak}
v_{peak} } Same units of	U
Q_{pl} } concentration	
Example 2(*b*)	
L	t_{peak}
U	Q (directly)
v at all t (arbitrary units)	

(3) *Using Some Properties of the Normal Error Curve*

The value of Pe calculated in the previous examples is relatively small and so the distribution of concentration is not symmetric enough for the method described in Section 3.4.2 to be applicable. But if readings were taken further downstream, then L, and hence Pe, would be larger, while the area under the curve will be the same. For example, if L were 117 ft and t_1, the time interval between the peak of the curve and the point where $v = 0.3679 v_{peak}$, is found to be 25.8 sec, then $|\theta_1| = t_1/(L/U) = 2.58 \times 10^{-1}$, and so, by Section 3.4.2, $\text{Pe} = 4/(|\theta_1|)^2 = 60.2$. Figure 3.8 shows the meaning of $|\theta_1|$. Figure 3.6 indicates that for this value of Pe the value of θ_{peak} approximates to unity; i.e., the curve is reasonably Gaussian in its central portion.

(4) *Statistical Calculation on the Concentrations Measured at Intervals*

The results are given as discrete values of the ordinate at intervals of 2 sec; if these are denoted by Δt, then

$$\begin{aligned}\sigma^2 &= [(\textstyle\sum vt^2\,\Delta t)/(\sum v\,\Delta t)] - [(\sum vt\,\Delta t)/(\sum \Delta t)]^2 \\ &= [(\textstyle\sum vt^2)/(\sum v)] - [(\sum vt)/(\sum v)]^2\end{aligned}$$

over all measured values. Any units can be used for v, while t can be measured from any origin: Levenspiel and Smith make the irruption of the source their origin of time, and so found the results given in Table 3.1. These yield

$$\begin{aligned}\textstyle\sum v &= 386 \quad \text{arbitrary units of concentration} \\ \textstyle\sum vt &= 4252 \quad \text{arbitrary units} \times \text{seconds} \\ \textstyle\sum vt^2 &= 65{,}392 \quad \text{arbitrary units} \times (\text{seconds})^2\end{aligned}$$

from which σ_t^2 was calculated as 48 sec². So, $\sigma_\theta^2 = 48/(L/U)^2 = 48/(7.7)^2 = 0.81$ in [reduced time (dimensionless) units]². And so from Equation (3.25) (rearranged) it follows that $\text{Pe} = 4.61$, as before.

3.6 THE USE OF A STEP CHANGE

The step change is simply related to the impulse as shown in texts on control or integral transforms. It is to be expected therefore that values of Pe and flow velocity could be obtained from this type of measurement. There is little to choose between the two methods from the point of view of analytical mathematics: practical considerations are of greater importance, and so are computational considerations.

3.6.1 Mathematical Form of a Degenerate Step Change

The equation

$$\frac{a\,\partial^2 v}{\partial z^2} - \frac{\partial v}{\partial t} = 0$$

subject to the conditions

$$\left.\begin{array}{ll} v = 1, & z > 0 \\ v = 0, & z < 0 \end{array}\right\} \quad t = 0; \qquad v \text{ is finite}, \quad z \longrightarrow \pm\infty, \quad \text{for all } t$$

has as its solution

$$v = \tfrac{1}{2}\{1 + \operatorname{erf}[z/2(\mathrm{at})^{1/2}]\} \qquad -\infty < z < \infty \tag{3.35}$$

(See Carslaw and Jaeger [7], Section 2.4, Equation 12.) In the above

$$\operatorname{erf}[z/2(\mathrm{at})^{1/2}] = (2/\pi^{1/2}) \int_0^{z/2(\mathrm{at})^{1/2}} \exp -d\xi^2 \, d\xi$$

z^2/at being the Fourier number. Equation (3.35) is the solution to the problem of finding the temperature of a slab, initially at unit temperature, and extending from zero to infinity, whose face at $z = 0$ is suddenly brought into contact with a similar semiinfinite slab at zero temperature. The shape of the distribution curve is shown in Figure 3.11, and shows how the shape alters as the time increases, the curves pivoting about $v = \frac{1}{2}$.

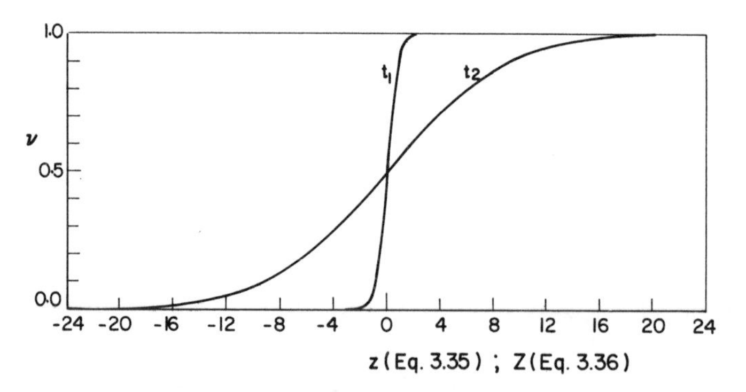

Fig. 3.11. Typical step change of concentration as function of Z; $t_2 = 100t_1$.

If the positions of the hot and cold slabs were to be reversed, then the equation for the distribution is $\nu = \frac{1}{2}\{1 - \text{erf}\,[z/2(at)^{1/2}]\} = \frac{1}{2}\{\text{erfc}[z/2(at)^{1/2}]\}$.

3.6.2 Step Change in a Moving Medium

If the initial discontinuity of Section 3.6.1 were to occur in a moving medium at the point $z = 0$, this discontinuity would travel at the velocity of the stream U; but if dispersive effects are present, then the discontinuity is transformed into a more gradual change as it drifts, the shape of the distribution being the same as in Figure 3.11 except that the abscissa is the distance Z measured from a moving origin. That is, $Z = z - Ut$ as in Chapter 2 and in Appendix 2. Hence, for a moving medium

$$\nu = \tfrac{1}{2}\{1 + \text{erf}[(z - Ut)/2(Dt)^{1/2}]\} \qquad (3.36)$$

The distribution of ν with Z has the shapes shown in Figure 3.11; i.e., the curves retain the symmetry of Figure 3.11 and ν is equal to $\frac{1}{2}$ at $z = Ut$. But if the distribution of ν with t is plotted, then this symmetry is lost; and Figure 3.12 shows this. In spite of this lack of symmetry, ν is still equal to

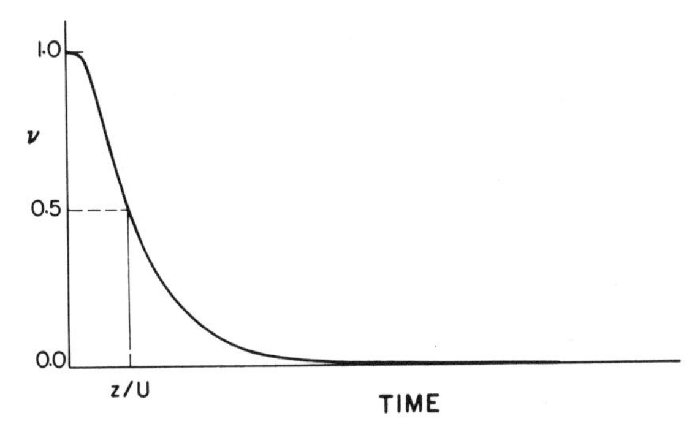

Fig. 3.12. Typical step change of concentration as a function of time.

$\frac{1}{2}$ at $t = z/u$ (for $\nu = \text{erf}\, Z = \text{erf}\, 0$), in agreement with the method of finding U that was described in Section 3.3; namely that of finding the time of mean area in the concentration–time curve of a degenerating impulse. For, the step change being in effect a succession of impulses, it follows that the distribution of a degenerating step change is the same as the integral of the degenerating impulse. In other words, the distribution caused by a step change is an analog of the integral of the distribution caused by an impulse; and so the integration has been done by physical means: one has to find, in place of an integral, the time at which the concentration rises to half its final value.

The variable in Equation (3.36) and hence the limits can be rearranged [as for Equation (3.9)] to give

$$\nu = \tfrac{1}{2}\left[1 + (2/\pi^{1/2}) \int_0^{[(1-\theta)/\theta^{1/2}](\text{Pe}^{1/2}/2)} \exp -\xi^2 \, d\xi\right]$$

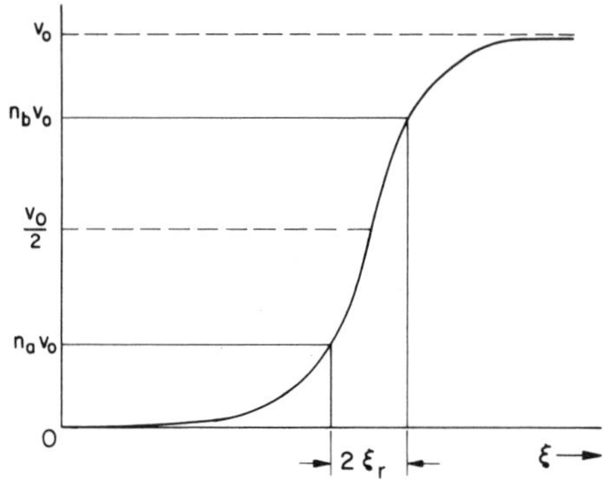

Fig. 3.13. Finding the value of Pe from a step change (after Hiby [8]).

or

$$\begin{aligned} \nu &= \tfrac{1}{2}\{1 + \operatorname{erf}[(1 - \theta)\,\mathrm{Pe}^{1/2}/2\theta^{1/2}]\} \\ &= \tfrac{1}{2}[1 + \operatorname{erf}(\mathrm{Pe}^{1/2}\Theta/2)] \end{aligned} \tag{3.37}$$

where $\Theta = (1 - \theta)/\theta^{1/2}$; this last form was used by Hiby [8, 9] to obtain a line (on probability paper) of ν versus Θ. If the diffusion model holds, then this line will be straight whatever the value of Pe, and so the validity of this model may be tested.

Example

Fluid flows at a velocity of 0.6 ft sec^{-1} through an infinite system; a step change was introduced and the concentration–time curve was determined at a point 7 ft downstream from the point where the step change occurred at $t = 0$. The relative concentrations were plotted on probability paper against Θ. It was found that the value of the time $t_{0.8}$ at which ν was 0.8 times the maximum was 8.7 sec. Hence, θ was $(8.7 \times 0.6)/7 = 0.746$ and $\Theta = 0.295$.

Hence,

$$0.8 = \tfrac{1}{2}[1 + \operatorname{erf}(0.295\mathrm{Pe}^{1/2}/2)]$$

i.e.,

$$0.6 = \operatorname{erf} 0.595 = \operatorname{erf} 0.1475\mathrm{Pe}^{1/2}$$

So,

$$\mathrm{Pe} = (0.595/0.1475)^2 = 16.24$$

and

$$D = UL/\mathrm{Pe} = (0.6 \times 7)/16.24 = 0.25 \quad \mathrm{ft^2\ sec^{-1}}$$

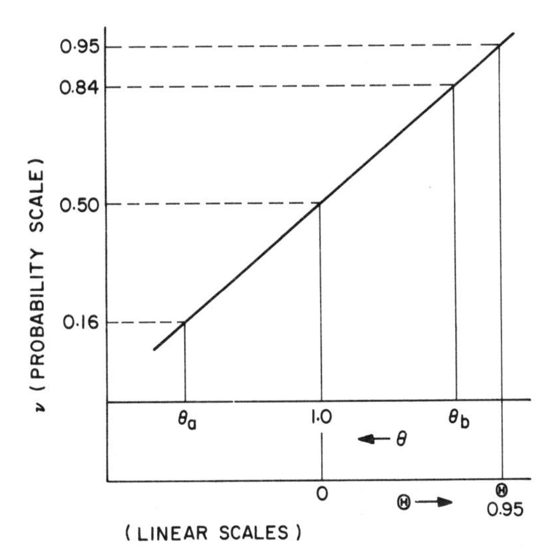

Fig. 3.14. Step change plotted on probability paper (after Hiby [8]).

3.6.3 Methods of Utilizing Analytic Properties of the Error Function Curve

It may be added that in the same way as the analytic properties of the normal curve were used in Section 3.4.2 to give a quick method of finding the value of σ^2 and hence of Pe, so the analytic properties of the integral of the normal curve should be expected to give a quick method of finding σ^2, and Appendix 4 shows that this is so. The experimental curve of concentration against time for large Pe or of concentration against Θ for all values of Pe is of the form of Figure 3.13, and in this the symbol ξ denotes either t or Θ as appropriate. This figure also shows the distance, designated by $2\xi_r$, between two given points on the curve; these distances are chosen to be certain fractions (n_p, $p = a, b$) of v_0, where v_0 is the asymptotically reached maximum concentration. As Appendix 4 shows, it is convenient to take these points such that $n_a = 0.16$ and $n_b = 0.84$. In this case $r = 2$ and $2\xi_2 = 2\sigma$ (in the units of ξ). [If $n_a = 0.08$ and $n_b = 0.92$, then Appendix 4 also shows that in this case $2\xi_1 = (2\sqrt{2})\sigma$, as obtained in Section 3.4.2.]

Since these last results can only be used for an error function, it follows that either the concentration must be plotted against Θ or, if plotted against θ, then Pe must be so large that the resultant curve is a close approximation to the error-function curve, at least in its central part.

Examples

In results due to Hiby [8] the value of Pe is so large that plotting the concentration against both Θ and θ will produce straight lines, as shown in Figure 3.14 (Hiby [8]). Hence, it should be possible, in this example, to use either method. By reading off values from the diagram, the following results were obtained.

Method 1

$$0.95 = \tfrac{1}{2}[1 + \text{erf}(\Theta \text{Pe}^{1/2}/2)]$$

$$\therefore \quad \text{erf}(\Theta \text{Pe}^{1/2}/2) = 0.9 = \text{erf}\, 1.163$$

Since $\Theta_{0.95} = 0.0555$, it follows that

$$\text{Pe} = (1.163 \times 2/0.0555)^2 = 1800 \quad \text{(approx.)}$$

Method 2

Since, in the example, only values of Θ are given in the diagram, it follows that the corresponding values of θ must be calculated. They are:

$$\theta_a = 1.033, \qquad \theta_b = 0.967$$

So, $2\sigma = 2\theta = 0.066$. Hence

$$\text{Pe} \doteqdot 2/\sigma^2 = 2/1.089 \times 10^{-3} = 1800 \quad \text{(approx.)}$$

as before; the uncertainty is due largely to the difficulty of taking off readings of Θ from the scales of the abscissa in Figure 3.14, which used Hiby's original results.

Two chapters having been devoted to discussing a situation that cannot arise in practice (because a true, or even a closely approximate, instantaneous source cannot be made), but which is often assumed to obtain, it is now proposed to examine methods that have been put forward to deal with pulses of practical shape. The next two chapters propound the arguments.

REFERENCES

[1] L. H. C. Tippett, "The Methods of Statistics." Benn, London, 1952.

[2] G. I. Taylor, Conditions under which Dispersion of a Solute in a Stream of Solvent can be used to Measure Molecular Diffusion. *Proc. Roy Soc.* (*London*) **A225**, 473 (1954).

[3] G. H. Hardy, "Pure Mathematics." Cambridge Univ. Press, London and New York, 1928.

[4] O. Levenspiel, "Chemical Reaction Engineering." Wiley, New York, 1971.

[5] O. Levenspiel, and W. K. Smith, Notes on the Diffusion-type Model for the Longitudinal Mixing in Flow. *Chem. Eng. Sci.* **6**, 227 (1957).

[6] C. E. Allen, and E. A. Taylor, The Salt Velocity Method of Water Measurement. *Trans. Amer. Soc. Mech. Eng.* **45**, 285 (1923).
[7] H.S. Carslaw, and J. C. Jaeger, "Conduction of Heat in Solids," 2nd. ed. Oxford Univ. Press, London and New York, 1959.
[8] J. W. Hiby, Third Congress of the European Fed. of Chem. Engrs. The Interaction between Fluids and Particles. *Inst. Chem. Eng. London* **C71** (1962).
[9] J. W. Hiby, *Chem.-Ing. Techn.* **30**, 180 (1958).

Chapter 4

The General Pulse; Properties and Uses of Its Transforms

4.1 INTRODUCTION

The difficulties of generating true impulses make it necessary to consider concentration waves of a more general shape and how they may be used to determine parameters in a flowing medium.

The simplest basic equation, obtained by quantity (mass or heat) balance over a differential length in the direction of motion, is, for a plane kinematic wave

$$\frac{\partial q_1}{\partial z} + \frac{\partial v}{\partial t} - q_j = 0 \qquad (6.2)\ \mathrm{R}$$

which is the second telegrapher's equation, referred to in Chapter 1 and derived in Section 6.2. In the above, q_1 is the longitudinal flux, as quantity/(time)·(unit cross-sectional area), and q_j is the flux from the jth stationary reservoir phase, as quantity/(time)·(unit volume of flowing medium). If q can be related to v by one or more linear equations, either algebraic or differential, then the kinematic wave may be called linear (Aris [1]).

It is now necessary to see whether such a general pulse can be used to give information about the system in which it travels. Since the parameters of the system will affect the shape of the pulse, it follows that if the pulse can be described mathematically in some way, it may be possible to determine these parameters if the proper measurements of the pulse shape can be made. There have been reported three methods of attack; all involve integral transforms of a humplike pulse as measured with time at one or two fixed locations.

4.2 THE USE OF THE MOMENTS OF THE PULSE

To summarize what follows, it may be said that the advantages of this procedure are that neither the shape of the initial pulse (e.g., Dirac impulse) nor an analytic equation of the wave is needed. On the other hand, when there are reservoir phases the algebraic expressions for the higher moments rapidly become very long and complicated and then the procedure does not seem capable of dealing with discontinuities in the path of the pulse. Furthermore, it is very sensitive to errors of measurement and to "noise"—i.e., random and uncertain fluctuations in the measured concentration.

4.2.1 Moments of a Pulse

The humplike wave of concentration dealt with here may be thought of as a statistical frequency or probability distribution. It may then be thought to possess moments, as explained in Appendix 7. These moments may be of the distribution against distance at a fixed time or of the distribution against time at a fixed distance. The latter will be considered here, but the former, experimentally difficult to determine, have been used with elegance in analytic investigations of the behavior of a medium toward a pulse; see Aris [2–4] and Levenspiel and Dayan [5, 6].

If the distribution of concentration against time is considered, the nth moment of it is defined by

$$m_n^* = \int_0^\infty t^n v(t, z)\, dt \tag{4.1a}$$

$$m_n = \int_0^\infty (t - m_1^*)^n v(t, z)\, dt \tag{4.1b}$$

Notes: (i) m^* denotes a moment about any origin, while m denotes a moment about the mean ($= m_1^*$); (ii) the moments must be finite for the present purpose; this requires that either $v = 0$ at finite t or that it approaches zero sufficiently quickly as $t \longrightarrow \infty$.

At any given value of z

$$m_0 = \int_0^\infty v\, dt \tag{4.2}$$

is proportional to the total amount of material in the pulse. If interchange with any reservoir phase is completely reversible, then m_0 is a constant and it may be normalized; that is, it may be put equal to unity by finding a constant C such that

$$C \int_0^\infty v\, dt \equiv \int_0^\infty (Cv)\, dt = 1$$

For example, if a pulse is generated by putting in 2.5 gm mol of solute into a stream of velocity 6 cm sec^{-1} and cross-sectional area 2 cm^2, then $2 \times 6 \int_0^\infty v\, dt = 2.5$, so C must be $2 \times 6/2.5 = 4.8$ $\text{cm}^3\ \text{sec}^{-1}\ \text{mol}^{-1}$ and the concentration at any time t (say 0.2 gm mole cm^{-3}) can be converted onto a new scale such that $m_0 = 1$. In the example the "concentration" will be $0.2 \times 4.8 = 0.96\ \text{sec}^{-1}$.

Again,

$$m_1^*/m_0 = \int_0^\infty vt\, dt \Big/ \int_0^\infty v\, dt = \bar{t} \tag{4.3}$$

where $\bar{t}$ is the mean time, i.e., the time interval between $t = 0$ and the first moment of area of the curve. This equals the mean residence time if certain conditions obtain.

As a special case the variance σ^2 is really a second moment about the mean and Appendix 7 shows that

$$\sigma^2 = (m_2^*/m_0) - (m_1^*/m_0)^2 \tag{4.4}$$

It should be noted that the dimensions of the nth moment are here $[\text{T}]^{n+1} \times$ [concentration], while those of the variance are $[\text{T}]^2$ and of m_n^*/m_0 are $[\text{T}]^n$.

The Laplace transform of $v(z, t)$ being

$$\bar{v}(z, s) = \int_0^\infty e^{-st} v(z, t)\, dt \tag{4.5}$$

it follows that

$$d\bar{v}/ds = -\int_0^\infty te^{-st}v\,dt \tag{4.6}$$

hence,

$$\lim(d\bar{v}/ds)_{s\to 0} = -\int_0^\infty tv\,dt = -m_1^* \tag{4.7}$$

i.e., the first moment, with the sign changed. A continuation of this process gives eventually

$$\lim(d^n\bar{v}/ds^n)_{s\to 0} = (-1)^n \int_0^\infty t^n v\,dt = (-1)^n m_n^* \tag{4.8}$$

(The conditions for this to hold are given by Aris [1].) Thus, if the equation of a wave is known only in the form of its Laplace transform (and this is often known, while its inversion is difficult and leads to a cumbersome expression), then expressions for its moments can be found by Equation (4.8); from these expressions some of the properties of the system can be deduced, as will now be exemplified, while Chapter 9 lists a few more examples.

4.2.2 Examples of the Use of Moments

Case 1: No Reservoir

Since $q_j = 0$, the equation is

$$D\frac{\partial^2 v}{\partial z^2} - U\frac{\partial v}{\partial z} - \frac{\partial v}{\partial t} = 0 \tag{4.9}$$

and the conditions are

$$\begin{aligned} v &= f(t), & z &= 0 \\ v &\to 0, & z &\to \infty \\ v &= 0, & t &= 0 \end{aligned}$$

The Laplace transform of Equation (4.9) gives a differential equation in z; the solution of this, along with the transform of the conditions, gives

$$\bar{v} = A(s)\exp[\chi \mathrm{Pe}\,\mu(s)] \equiv A(s)\exp[Uz\mu(s)/D] \tag{4.10}$$

where

$$\chi = z/L \tag{4.11}$$

$$\mu = \tfrac{1}{2} - p(s) \tag{4.12}$$

$$p(s) = [\tfrac{1}{4} + (Ds/U^2)]^{1/2} \tag{4.13}$$

Here, $A(s, 0)$ is a constant with respect to χ [it is the Laplace transform of $v(t, 0)$ and its value is not known], and L is a reference length in the direction of flow. If primes on A denote differentiation with respect to s, and the subscript 0 indicates the limiting value as $s \to 0$, then application of the process in Equation (4.8) to Equation (4.10) gives

$$\bar{v}_0 \equiv m_0 = A(0) \tag{4.14}$$

$$-\bar{v}_0' \equiv m_1^* = -\{A'(0) - [A(0)z/U]\} \tag{4.15}$$

$$\bar{v}_0'' \equiv m_2^* = A''(0) - A'(0)(2z/U) + A(0)[(z/U)^2 + (2zD/U^3)] \tag{4.16}$$

At two points in the stream, viz., a (where $z = 0$) and b (where $z = L$), the use of Equations (4.14)–(4.16) gives, by elimination of A, A', and A'',

$$(m_1^*/m_0)_b - (m_1^*/m_0)_a = L/U \equiv t_0 \tag{4.17}$$

$$\begin{aligned}\sigma_b^2 - \sigma_a^2 &\equiv \{[m_0 m_2^* - (m_1^*)^2]/m_0^2\}_b \\ &\quad - \{[m_0 m_2^* - (m_1^*)^2]/m_0^2\}_a \\ &= 2LD/U^3 \\ &\equiv 2t_0^2/\mathrm{Pe}\end{aligned} \Bigg\} \; [\mathrm{T}^2] \tag{4.18}$$

and (Chao and Hoelscher [7])

$$\pi_b^3 - \pi_a^3 = 12t_0^3/\mathrm{Pe}^2 \tag{4.19}$$

where $\pi^3 = m_3/m_0$ is the third normalized moment about the mean. These results show two things. First, for a pulse governed by Equation (4.9) the first moment of area moves with the speed of the stream. A pulse, being of ever-changing shape, has no very obvious speed, but it could well be defined as $V_w = z/\tilde{m}_1$, where $\tilde{m}_1$ is the difference between the values of (m_1^*/m_0) for the wave at two points which are a distance z apart. Hence, from Equation (4.17) it is seen that $V_w = U$ and so it does not alter with time if U is constant. (This has already been shown in Chapter 3 for a perfect impulse.)

Second, Equation (4.18) shows that a measure of D can be obtained by determining the difference between the variances of the wave at two points, again at a known distance apart, in a stream of known velocity. Alternatively, L can be eliminated between Equations (4.17) and (4.18) to give

$$D = \frac{U^2(\sigma_b^2 - \sigma_a^2)}{2[(m_1^*/m_0)_b - (m_1^*/m_0)_a]} \tag{4.20}$$

An example of finding both the variance of a concentration–time curve and of m_1^*/m_0 will be found in Chapter 3. Typical units used in Equation (4.20) would be: numerator, (cm/sec)^2 sec^2 and denominator, (concentration $\times$ sec^2)/(concentration $\times$ sec). Equation (4.19) would also give Pe, but third moments would be even more sensitive to errors of measurement. The situation is summarized in Figure 4.1.

Since the above conclusion is a general one, it must apply to the special case of a Gaussian concentration–time curve. It can be shown (e.g., Aris [1, 2]) that linear pulses of the present type tend ultimately to become Gaussian with respect to t, at least in the central part. Waves of this shape, having a simple analytic form, allow of the easy way of finding the variance that was described in Chapter 3 and in Appendix 4. Hence, as shown in Figure 4.2, if a wave of arbitrary shape is injected sufficiently far upstream (at point b), it will have attained a Gaussian shape (with respect to t) by the time

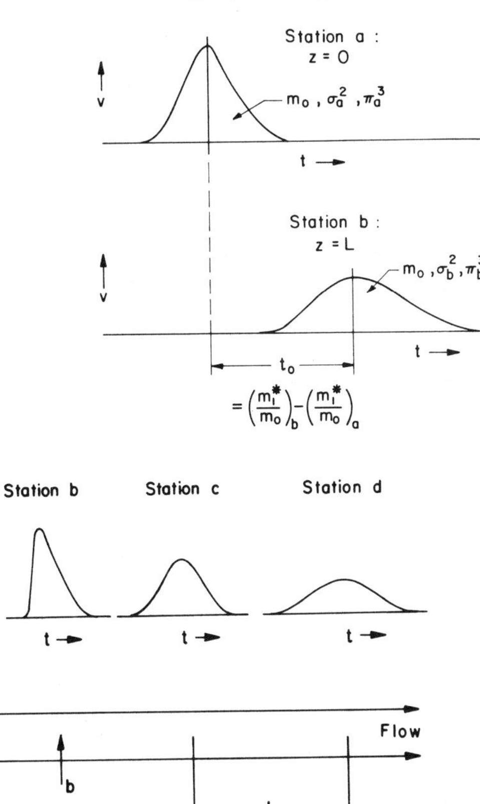

Fig. 4.1. Change of moments as pulse progresses.

Fig. 4.2. Gaussian concentration curves.

it reaches the two measuring stations c and d a distance L apart—the only distance required to be known. To observers at the latter two stations the wave appears to have had origin as a perfect impulse at point a. Hence, substitution of the symbols for distances from Figure 4.2 into Equation (3.25) gives

$$\sigma_c^2 U^2/z^2 = (8D^2 + 2DUz)/z^2U^2$$

and

$$\sigma_d^2 U^2/(z + L)^2 = [8D^2 + 2DU(z + L)]/(z + L)^2U^2$$

So, $D = (\sigma_d^2 - \sigma_c^2)U^3/2L$, as already shown in this chapter, but now σ^2 can be readily found. [An example of units is: (sec^2 — sec^2) $(\text{cm/sec})^3$/cm.]

Case 2: Reservoir Phases Present

The wave equation is now

$$D\frac{\partial^2 v}{\partial z^2} - U\frac{\partial v}{\partial z} - \frac{\partial v}{\partial t} + \sum_{j=2}^{j} q_j = 0 \tag{4.21}$$

in which $q_j(t)$ is the flux from the jth reservoir phase, as quantity per unit time to unit volume of flowing phase. That is, it is the rate at any instant across an interface separating the jth fixed reservoir phase from the flowing phase. As written, the equation implies that the fluxes are *in parallel*, i.e., acting independently of one another at any value of t and z. It does not say anything about the detailed structure of the reservoir phases supplying these fluxes; this structure may involve fluxes *in series*. The italicized words are taken from electrical terminology and are illustrated in Figure 4.3.

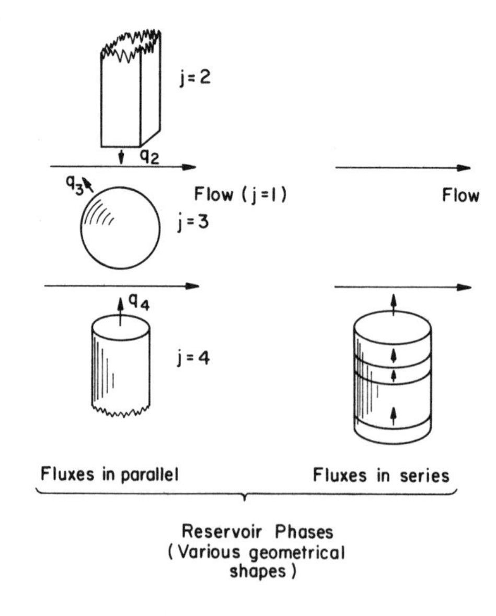

Fig. 4.3. Reservoir phases in parallel and series (illustrative).

In order to proceed, it is necessary to express the flux as a function of the concentration in the flowing phase; i.e.,

$$q_j(z, t) = F_{1j}[v(z, t), v_{\text{surf}}, \text{parameters of reservoir}] \tag{4.22}$$

where v_{surf} is the concentration at the interface surface of the reservoir phase. Then, substitution of the Laplace transform of this last relation into the transform of Equation (4.21) gives

$$D\frac{\partial^2 \bar{v}(z, s)}{\partial z^2} - U\frac{\partial \bar{v}(z, s)}{\partial z} - [s\bar{v}(z, s) - v(z, t = 0)]$$
$$+ \sum F_{1j}[\bar{v}(z, s), \bar{v}_{surf}(z, s), \text{etc.}] \tag{4.23}$$

The simplification is usually made that $v(z, t = 0) = 0$. It is necessary to express $\bar{v}_{surf}(z, s)$ as a function of $\bar{v}(s, z)$; i.e.,

$$\bar{v}_{surf} = F_2(\bar{v}) \tag{4.24}$$

[This is obtained from the differential equation of the reservoir phase and boundary conditions; from these (in the Laplace domain), expressions for the flux at, and the concentration of, the reservoir interface are obtained in terms of $\bar{v}$.] Thus by, using Equations (4.23) and (4.24), an equation results of the form

$$D\frac{\partial^2 \bar{v}}{\partial z^2} - U\frac{\partial \bar{v}}{\partial z} + F_3(\bar{v}) = 0 \tag{4.25}$$

This, if solved in transformed space subject to the conditions that

(a) $\lim(\bar{v})_{z\to\infty} =$ finite;
(b) $\bar{v}(0, s) = A(s)$; i.e., $v(0, t)$ is a continuous but unknown function of time, and $A(s)$ is unknown;
(c) $\bar{v}(z, 0) = 0$;

will give generally as a solution

$$\bar{v} = F_4(z, s) \tag{4.26}$$

This solution, for the case where

$$F_3(\bar{v}) = -F_5(s) \cdot \bar{v}(z, s) \tag{4.27}$$

is

$$\bar{v} \equiv F_4(z, s) = A(s)\{\exp[(U/2D) - [(U/2D)^2 + F_5(s)]^{1/2}]z\} \tag{4.28}$$

Then the nth moment is obtained as in Equation (4.8).

Examples (*See also Chapter 9.*)

EXAMPLE 1 $q_j = 0$

This is now seen to be a special case of Equation (4.27); it was treated as Case 1, above.

EXAMPLE 2

Single reservoir phase of infinite conductivity, finite external resistance [7]. See Section 7.3 and Figure 7.1.

Now,

$$q_2 = -\tilde{V}_2\, \partial v_2/\partial t \tag{7.2) R$$

and

$$\partial v_2/\partial t = \kappa_1 v_1 - \kappa_2 v_2 \tag{7.7 R}$$

The Laplace transform of the last equation gives

$$\bar{v}_2(s) = [\kappa_1/(s + \kappa_2)]\bar{v}_1(s) \tag{4.29}$$

hence,

$$\bar{q}_2(s) = (-\tilde{V}_2 s)\bar{v}_2(s) = [-\tilde{V}_2\kappa_1 s/(s + \kappa_2)]\bar{v}_1(s) \tag{4.30}$$

and so

$$F_5(s) = s + [\tilde{V}_2\kappa_1 s/(s + \kappa_2)] \tag{4.31}$$

which can be substituted into Equation (4.28) and the differences in the moments of the pulse at two measuring stations obtained (see Aris [8] and Chao and Hoelscher [7]). The expressions contain the parameters $\tilde{V}$, $\mathbf{K}$, U, D, $\tilde{a}$, k, and L. If $\tilde{V}$ is known, the parameters can be grouped as

$$U/k\tilde{a}L \quad \text{and} \quad UL/D$$

Thus, these two groups, as well as $\mathbf{K}$, can be found from two moments even if the values of the individual parameters cannot.

Example 3. Resistances in Series

A model due to Kučera in which several different mechanisms govern the transfer of material in series is discussed in Section 7.5.

4.3 CALCULATIONS IN THE LAPLACE DOMAIN

The large errors that can arise from use of moments may perhaps be reduced by finding the numerical values of the Laplace transforms of the concentration–time distributions at $z = 0$ and $z = L$; (perhaps a slightly unusual use of the transform). It has been used by Østergaard and Michelson [9] and Anderssen and White [15]. Actually, the finite transform is used, just as the finite Fourier transform is used in Section 4.4, and for the same reason; viz., to limit the amount of computation. However, it can usually be arranged that the time interval can be both practical and sufficient. Thus, now,

$$\bar{v}(s) = \int_0^T v(t)e^{-st}\,dt/m_0 \tag{4.32}$$

where $m_0 = \int_0^\infty v(t)\,dt \propto$ total tracer amount; it could be measured separately, but is not usually needed because it is the same for both $z = 0$ (station a) and $z = L$ (station b).

The ratio of the Laplace transforms at inlet and outlet [found from Equation (4.32)] is computed for various numerical values of s in any units (for example, $\sec^{-1}$ could be used). Thus,

$$F(s) = [\bar{v}(s)]_b/[\bar{v}(s)]_a \tag{4.33}$$

The analytic expression for $F(s)$ now needs to be known; it will depend, of course, on the differential equation describing the system.

For example, if $q_j = 0$, then the appropriate equation is (4.10), and so from that there arises

$$\begin{aligned} F(s) &= \exp[\mathrm{Pe}\mu(s)]1 \\ &\equiv \exp(\tfrac{1}{2}\mathrm{Pe}\{1 - [1 + (4st_0/\mathrm{Pe})]^{1/2}\}) \end{aligned} \tag{4.34}$$

This may be rewritten as

$$(\ln\{[F(s)]^{-1}\})^{-1} = st_0(\ln\{[F(s)]^{-1}\})^{-2} - \mathrm{Pe}^{-1} \tag{4.35}$$

or

$$Y(s) = t_0 X(s) - \mathrm{Pe}^{-1} \tag{4.36}$$

and so a plot of Y versus X, using the experimentally computed values of $F(s)$, would give a straight line of intercept $-\mathrm{Pe}^{-1}$ and of slope t_0 (i.e., L/U).

An alternative method has been suggested by Østergaard and Michelsen. For, $F = \bar{v}_o/\bar{v}_i$. Thus,

$$\frac{1}{F}\frac{dF}{ds} = \frac{1}{\bar{v}_o}\frac{d\bar{v}_o}{ds} - \frac{1}{\bar{v}_i}\frac{d\bar{v}_i}{ds} \tag{4.37}$$

$$\frac{d}{ds}\left(\frac{1}{F}\frac{dF}{ds}\right) = \left[\frac{1}{\bar{v}_o}\frac{d^2\bar{v}_o}{ds^2} - \frac{1}{\bar{v}_o^{\,2}}\left(\frac{d\bar{v}_o}{ds}\right)^2\right] - \left[\frac{1}{\bar{v}_i}\frac{d^2\bar{v}_i}{ds^2} - \frac{1}{\bar{v}_i^{\,2}}\left(\frac{d\bar{v}_i}{ds}\right)^2\right] \tag{4.38}$$

Now, if

$$\bar{\mu}_n = \int_0^\infty (t - m_1^*)^n v e^{-st}\, dt/\bar{\mu}_0 \tag{4.39}$$

$$\bar{\mu}_1 = \int_0^\infty t v e^{-st}\, dt/\bar{\mu}_0 \tag{4.40}$$

and

$$\bar{\mu}_0 = \int_0^\infty v e^{-st}\, dt \equiv \bar{v}(s) \tag{4.41}$$

then

$$\begin{aligned} d\bar{v}/ds &= (d/ds)\left(\int_0^\infty v e^{-st}\, dt\right) \\ &= -\int_0^\infty t v e^{-st}\, dt \\ &= -\bar{\mu}_1\bar{\mu}_0 \end{aligned} \tag{4.42}$$

and similarly,

$$d^2\bar{v}/ds^2 = \bar{\mu}_0(\bar{\mu}_2 + \bar{\mu}_1^{\,2}) \tag{4.43}$$

So, Equation (4.38) becomes

$$\frac{d}{ds}\left(\frac{1}{F}\frac{dF}{ds}\right) = (\bar{\mu}_2)_o - (\bar{\mu}_2)_i \tag{4.44}$$

By an extension of the above argument, an expression for

$$\frac{d^n}{ds^n}\left(\frac{1}{F}\frac{dF}{ds}\right)$$

can be found in terms of the transforms of the first, second, . . . , nth moments. If the differential equation of the system is known, the expressions for these transformed moments may be found.

Thus, to use the same example as before (viz., $q_j = 0$), it is found that

$$\frac{d^n}{ds^n}\left(\frac{1}{F}\frac{dF}{ds}\right) = (-t_0)^{n+1}\text{Pe}^{-n}\frac{(2n)!}{n!}\left(1 + \frac{4st_0}{\text{Pe}}\right)^{-[n+(1/2)]} \tag{4.45}$$

Additional parameters in the system will of course give rise to more complicated expressions for the transformed moments. If there are n parameters, these could perhaps be found by determining n different moments at one value of s, or one moment at n values of s.

4.4 THE USE OF THE FOURIER TRANSFORM

Harmonic analysis of periodic, nonsinusoidal waves may be used to extract both fundamental and harmonics, and these latter have been used [10] to determine parameters in the way outlined elsewhere in this book for frequency response.

However, the extraction of these sinusoidal components from a pulse as opposed to repetitive waves is interesting in that it provides a link between the use of a pulse and of frequency response for parameter determination.

This use of a pulse for determining frequency response has been widely used in investigations of the behavior and control of chemical plant, mechanical and aeronautical systems, and electrical circuits; its interest there lies largely in the fact that the analytical power of measurements in the frequency domain is needed but the disturbance which an actual system would suffer by being continuously perturbed would not be tolerable for practical or commercial reasons.

In the laboratory these latter considerations have less or no weight, so the determination of parameters by utilizing frequency response obtained by a pulse would need justification for another reason. This would seem to be the consequent lack of need of a sine-wave generator, while at the same time the experimenter enjoys the advantages of frequency-domain evaluations which are listed in Chapter 9 and discussed in detail in Chapters 6, 7, and 8. However, the method possesses several sources of error which are not necessarily obvious; these are summarized below.

The reverse process—obtaining the impulse response from frequency response—is also possible.

4.4.1 Basic Procedure

If a pulse of concentration, exemplified by Figure 4.4, is $v(t)$, and is of duration T_w (also shown), then the frequency content of the pulse $S(\omega)$ is

$$S(\omega) = F[v(t)] = \int_{-\infty}^{\infty} v(t)e^{-i\omega t}\, dt = \int_0^T v(t)e^{-i\omega t}\, dt \tag{4.46}$$

where F denotes Fourier transform. $S(\omega)$ is continuous in ω under certain restrictions; see Sneddon [11]. It can be normalized by dividing by

$$S(0) = \int_0^T v(t)e^0\, dt = m_0$$

but usually the response, viz.,

$$G(i\omega) = F[v(t)_{\text{out}}]/F[v(t)_{\text{in}}] \tag{4.47}$$

is required, so normalization is then not necessary.

By the Euler relationship

$$e^{-i\omega t} = \cos \omega t - i \sin \omega t$$

and from the definitions

$$A = \int_0^T v(t) \cos \omega t\, dt \tag{4.48}$$

$$B = \int_0^T v(t) \sin \omega t\, dt \tag{4.49}$$

the ratio of the amplitudes (the magnitude ratio) of the output and input sine wave of frequency ω is

$$\text{M.R.}(\omega) = (\mathcal{R}^2 + \mathcal{I}^2)^{1/2} \tag{4.50}$$

and the phase lag ψ is

$$\psi = \tan^{-1}[\mathcal{I}/\mathcal{R}] \tag{4.51}$$

where

$$\mathcal{R} = (A_o A_i + B_o B_i)/(A_i^2 + B_i^2) \tag{4.52}$$

$$\mathcal{I} = (A_o B_i - B_o A_i)/(A_i^2 + B_i^2) \tag{4.53}$$

subscripts o and i referring to output and input, respectively.

In principle, then, the response at any frequency ω may be found by finding the integrals A and B at the specified value of ω from the experimental curve of $v(t)$ against t. This, then, is a problem in numerical methods.

4.4.2 Accuracy of the Determinations

If, say, the trapezoidal rule or Simpson's rule is used as a quadrature formula, the fact that $\cos \omega t$ and $\sin \omega t$ are involved means that the product curves, viz., $v(t) \cos \omega t$ and $v(t) \sin \omega t$, oscillate faster and faster as the value of ω is increased, and special methods have to be used, otherwise, above a certain frequency, the quadrature formula becomes useless. Clements and Schnelle [12], Dreifke [13], and Hougen and Walsh [14] discuss this. Clements and Schnelle [12] also contain an account of other possible sources of error. These include:

(i) The effect of the accuracy with which $v(t)$ is known.

(ii) The effect of truncation [because $v(t)$ is not exactly zero at $t \geq T$].

(iii) The effect of too large a time interval. [To save computational time, the time intervals Δt must not be too small. Their magnitude can depend on the rate of change of $v(t)$, and two or more intervals may be used, as illustrated by Δt_1 and Δt_2 in Figure 4.4.]

(iv) The frequency content of the pulse (perhaps the most important source of error). It is usually true that the frequencies which have to be used to get information are high in the sense that they are relatively difficult to generate or give rise to computational or measurement errors. It might be thought that the use of the Fourier transform of a pulse might avoid the difficulties of generating continuous high-frequency waves, but this process has its own difficulties. Thus, first, the frequency responses of the input and output detecting elements both have to be good, and equal, at high frequencies, whatever method is used, for otherwise, only a highly attenuated final signal is obtained. Second, the oscillation of the product curves, mentioned above, has to be dealt with. Third, just as in methods of using pulses, the ideal—aimed for but unrealizable—is a Dirac impulse. In the present case its transform contains all frequencies and of equal amplitudes, so the natural tendency is to strive to approach this as closely as possible. The outcome is an approximation to a square pulse, of finite width. Now, it so happens that this is one of the worst-shaped pulses to be used, for (a) the amplitudes of the higher harmonics decrease more rapidly than from other pulses, and (b) while many other pulses have zero content at certain frequencies, for a square pulse the first zero is reached at a lower frequency than others, as shown in Figure 4.5. [In remarks (a) and (b) there is a condition that differently shaped pulses have the same pulse width (T_w in Figure 4.4).] Clements and Schnelle [12] advise that the frequencies used should be less than that of the first zero

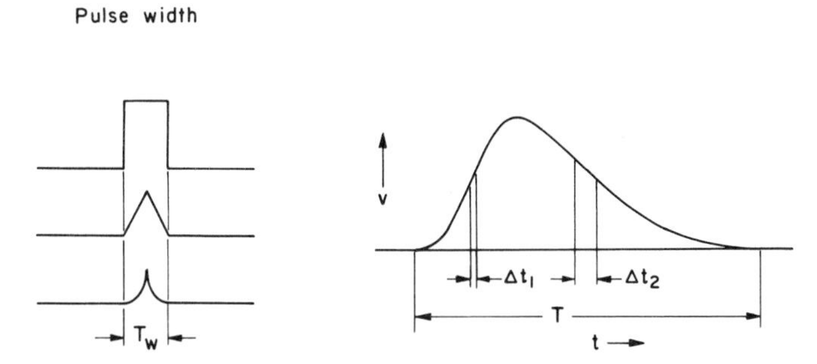

Fig. 4.4. Fourier transform from a pulse of width T_w.

content. They also point out that if this frequency were inadvertently used, the magnitude ratio and phase angle should theoretically be indeterminate (0/0), but due to small observational errors, finite but highly inaccurate values may be computed at this frequency.

The net practical consequences of these features would be for the experimenter to strive to produce a pulse that is both of short duration and of ordained shape in order that the content of the higher harmonics should be

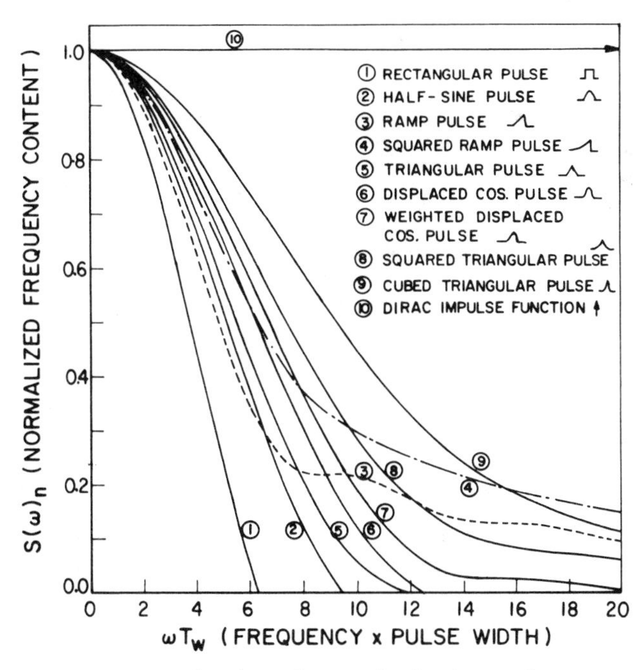

Fig. 4.5. Frequency content of various shapes of pulse (up to first zero only). (With acknowledgments to Clements and Schnelle [12].)

relatively large, and of significant intensity in order that this content be absolutely large. The consequences of the latter quality being the serious risk of nonlinearity and breakdown of the model, and the demands of the first-mentioned specifications requiring much time and ingenuity, it is suggested that in the end there is no practical advantage over the use of a steady-state sinusoidal wave.

It was assumed above that input and output pulses (v_i and v_o) would be measured. If only the output is measured, the situation is much worse, for instead of the assumed uniform frequency content of, say, a Dirac impulse, the actual input would be the unknown content of, say, an imperfect square pulse.

REFERENCES

[1] R. Aris, On the Dispersion of Linear Kinematic Waves. *Proc. Roy. Soc. (London)* **A245**, 268–277 (1958).
[2] R. Aris, On the Dispersion of a Solute in a Fluid Flowing through a Tube. *Proc. Roy. Soc. (London)* **A235**, 67–77 (1956).
[3] R. Aris, On the Dispersion of a Solute by Diffusion, Convection and Exchange between Phases. *Proc. Roy. Soc. (London)* **A252**, 538–550 (1959).
[4] R. Aris, The Longitudinal Diffusion Coefficient in Flow through a Tube with Stagnant Pockets. *Chem. Eng. Sci.* **11**, 194–198 (1959).
[5] J. Dayan and O. Levenspiel, Longitudinal Dispersion in Packed Beds of Porous Adsorbing Solids. *Chem. Eng. Sci.* **23**, 1327–1334 (1968).
[6] J. Dayan and O. Levenspiel, Dispersion in Smooth Pipes with Adsorbing Walls. *Ind. Eng. Chem. Fundam.* **8**, 840–842 (1969).
[7] R. Chao and H. E. Hoelscher, Simultaneous Axial Dispersion and Adsorption in Packed Beds. *AIChE J.* **12**, 271–278 (1966).
[8] R. Aris, On Shape Factors for Irregular Particles—II The Transient Problem. Heat Transfer in a Packed Bed. *Chem. Eng. Sci.* **7**, 8–14 (1957).
[9] K. Østergaard and M. L. Michelsen, On the Use of the Imperfect Tracer Pulse Method for Determination of Hold-up and Axial Mixing. *Canad. J. Chem. Eng.* **47**, 107–112 (1969).
[10] H. Littman and A. P. Stone, Gas-Particle Heat Transfer Coefficients in Fluidized Beds by Frequency Response Techniques. *57th Ann. Meeting, Amer. Inst. Chem. Eng., Boston* Dec. 6–10 (1964); *Chem. Eng. Progr. Symp. Ser.* **62,** No. 62 (1966).
[11] I. N. Sneddon, "Fourier Transforms." McGraw-Hill, New York, 1951.
[12] W. C. Clements, Jr., and K. B. Schnelle, Jr., Pulse Testing for Dynamic Analysis. *Ind. Eng. Chem., Process Design Develop.* **2**, 94–102 (1963).
[13] G. E. Dreifke, Effects of Input Pulse Shape and Width on Accuracy of Dynamic System Analysis from Experimental Pulse Data. Sc.D. Thesis, Washington Univ. (1961).
[14] J. O. Hougen and R. A. Walsh, Pulse Testing Method. *Chem. Eng. Prog.* **57**, 69 (1961).
[15] A. S. Anderssen and E. T. White, Parameter Estimation by the Weighted Moments Method. *Chem. Eng. Sci.* **26**, 1203–1221 (1971).

Chapter 5

Behavior at a Longitudinal Discontinuity: Pulses

5.1 BOUNDARY CONDITIONS

In Figure 5.1(a) a pulse or wave is traveling in either direction across the discontinuity shown. The conditions to be satisfied have (at least in the case of a steady-state chemical reactor) been the subject of some debate and discussion [1–5]. In particular, Wehner and Wilhelm [5] have summarized the situation, both for a reactor and, in an appendix, for a time-varying, nonreacting system, and Bischoff [6] extended their analysis to higher-order kinetics.

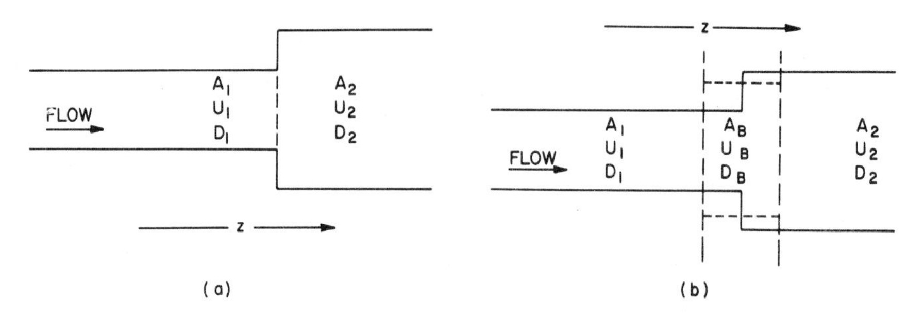

Fig. 5.1. Conditions at a longitudinal boundary: (a) idealized case; (b) an approximation to the actual situation.

The basic conditions to be satisfied at the boundary are:

(i) There is no discontinuity of concentration, hence

$$v_1 = v_2 \tag{5.1}$$

(See, however, Chapter 1 for comments.)

(ii) There is no accumulation of material (or heat); i.e., the total amount per unit time brought up to the boundary by convection and diffusion is equal to that leaving the boundary by the same processes; i.e.,

$$A_1 U_1 v_1 - A_1 D_1 \,\partial v_1/\partial z = A_2 U_2 v_2 - A_2 D_2 \,\partial v_2/\partial z \tag{5.2}$$

the partial derivative being necessary because the statement is true for all time; Wehner and Wilhelm's equations, referring to steady-state reactors, contain ordinary differential coefficients.

By comparing Equation (5.2) and the so-called first telegrapher's equation (6.1a), it will be seen that Equation (5.2) could be written as

$$A_1 q_1 = A_2 q_2 \tag{5.3}$$

where q is a longitudinal flux per unit cross-sectional area.

Thus, Equations (5.1) and (5.2) are analogous to the conditions which describe the behavior at a junction in an electric network or transmission-line system, v being the analog of voltage and Aq that of current.

Again, the equation of continuity of mass flow gives

$$A_1 U_1 \rho_1 = A_2 U_2 \rho_2 \tag{5.4}$$

If $\rho_1 = \rho_2$, then the last equation may be substituted into Equation (5.2) to give

$$v_1 - (D_1/U_1) \,\partial v_1/\partial z = v_2 - (D_2/U_2) \,\partial v_2/\partial z \tag{5.5}$$

Finally, condition (5.1) applied to the last equation gives

$$(D_1/U_1)\,\partial v_1/\partial z = (D_2/U_2)\,\partial v_2/\partial z \qquad (5.6)$$

That is, the ratio of the gradients across the boundary is a constant for all time, the gradient in a section being inversely proportional to D/U (or proportional to the Peclet number UL/D, where L is any reference length) in its respective section.

In an actual situation, especially where there is a change of area, i.e., of hydrodynamic regime, the situation is nearer to Figure 5.1(b), where in the boundary region, of ill-defined length, the values of U_B and D_B would probably a different from those obtaining in either region, and would probably be functions of z. Fan and Ahn [3] have reported on an experimental evaluation of boundary conditions in a steady-state reactor.

In a steady-state flow reactor the "Danckwerts boundary condition" [2], viz., that $dv/dz = 0$ at the exit, was shown by Wehner and Wilhelm [5] to follow from their general conditions—Equations (5.1) and (5.2) with an ordinary differential coefficient. It has been further discussed by Pearson [4]. It has been applied to problems where time-varying concentrations occur in nonreacting systems, but this would appear unnecessarily restrictive. Although somewhat out of context of this book, the "Danckwerts condition" for a reactor is discussed again at the end of this chapter because of the confusion that arises over its use in a nonreacting system.

The remainder of the chapter deals with pulses at an ideal boundary (Figure 5.1a), while in Chapter 6 the behavior of sine waves at such a boundary, including the concept of reflections and the ease with which sine waves overcome the problems of incorporating boundary conditions into an experimental determination, are discussed; in particular the "Danckwerts condition" for sine waves is discussed in Section 6.10.

5.2 THE MOMENTS OF A PULSE: BOUNDARIES NOT AT INFINITY

Chapters 2 and 3 dealt with the special case of a perfect impulse in one-dimensional flow with boundaries at infinity. Chapter 4 introduced the idea of computing the transforms of a disturbance, again with boundaries at infinity. But when the boundary conditions are those of Section 5.1, the winning of a solution, by classical methods, that satisfies all the conditions is much more difficult. The use of moments can deal with more demanding boundary conditions in an elegant way, at least in simple systems without reservoir phases. A perfect impulse will first be considered, but later a generalized wave will be dealt with, surging in a system with a number of boundary conditions, but again without reservoir phases. Analytic expressions for the mean residence time and the variance of the concentration–time curve are obtained (as functions of Pe and other parameters) without the necessity for inverting the Laplace transforms.

5.2.1 Perfect Impulse: Two Boundaries

The system is shown in Figure 5.2, the central region being bounded. Following van der Laan [7], the differential equations are written in terms of quantities which are both dimensionless and normalized.

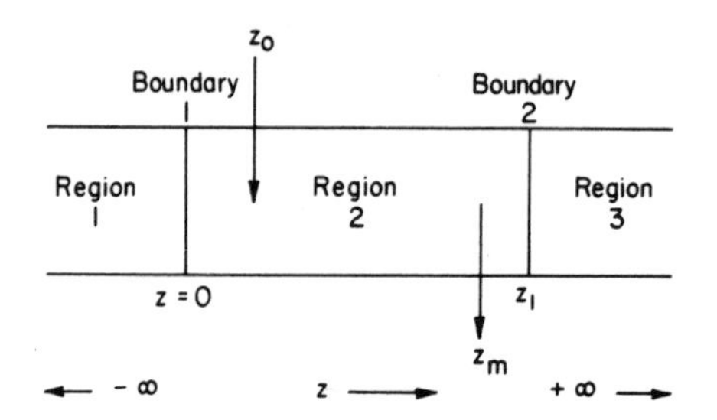

Fig. 5.2. Two longitudinal boundaries; moments of an impulse at one observation point.

Thus, for the single-phase system of Figure 5.2 in which an impulse is put in at z_0 at time zero and detected at z_{m}, the equations are

$$\left(\frac{1}{\mathrm{Pe}_n}\right)\frac{\partial^2 \mathcal{R}_n}{\partial \chi^2} - \frac{\partial \mathcal{R}_n}{\partial \chi} - \frac{\partial \mathcal{R}_n}{\partial \theta} = 0, \qquad n = 1, 3 \tag{5.7}$$

but in region 2, where the source irrupts,

$$\left(\frac{1}{\mathrm{Pe}_2}\right)\frac{\partial^2 \mathcal{R}_2}{\partial \chi^2} - \frac{\partial \mathcal{R}_2}{\partial \chi} - \frac{\partial \mathcal{R}_2}{\partial \theta} = \delta(\theta)\,\delta(\chi - \chi_0) \tag{5.8}$$

where $\mathcal{R} = Mv/\tilde{M}$; it is such that

$$\tilde{M} = \int_{-\infty}^{\infty} v\,dz = U\int_0^{\infty} vL(dt/L) = \int_0^{\infty} vM\,d\theta$$

i.e.,

$$\int_0^{\infty} \mathcal{R}\,d\theta \equiv m_0 = 1$$

with M the volume of fluid of a concentration (taken as unity) that would contain the same amount of solute as did the pulse in a conduit of unit cross-sectional area; v is a dimensionless concentration, equal to the number of times that a concentration is stronger than the one taken as unity;

$$\chi = z/L$$
$$\theta = tU/L = t/t_0, \qquad \text{where} \quad t_0 = L/U$$
$$L = z_{\mathrm{m}} - z_0$$

and δ denotes the Dirac delta function, here applying to both time and to distance.

The conditions are

$$\mathcal{R}_1 = \mathcal{R}_2 = \mathcal{R}_3 = 0, \qquad \theta \leq 0 \tag{5.9}$$

$$\mathcal{R}_1 \quad \text{is finite as} \quad \chi \longrightarrow -\infty \tag{5.10}$$

$$\mathcal{R}_3 \quad \text{is finite as} \quad \chi \longrightarrow +\infty \tag{5.11}$$

$$\mathcal{R}_1(0^-) - (1/\text{Pe}_1)\,\partial\mathcal{R}_1(0^-)/\partial\chi = \mathcal{R}_2(0^+) - (1/\text{Pe}_2)\,\partial\mathcal{R}_2(0^+)/\partial\chi \tag{5.12}$$

$$\mathcal{R}_2(\chi_1^{\,-}) - (1/\text{Pe}_2)\,\partial\mathcal{R}_2(\chi_1^{\,-})/\partial\chi = \mathcal{R}_3(\chi_1^{\,+}) - (1/\text{Pe}_3)\,\partial\mathcal{R}_3(\chi_1^{\,+})/\partial\chi \tag{5.13}$$

$$\mathcal{R}_1(0^-) = \mathcal{R}_2(0^+) \tag{5.14}$$

$$\mathcal{R}_2(\chi_1^{\,-}) = \mathcal{R}_3(\chi_1^{\,+}) \tag{5.15}$$

The last four conditions are those of Section 5.1 in dimensionless form. The Laplace transformations of Equations (5.7) and (5.9) lead immediately to

$$\bar{\mathcal{R}}_1 = A_1(\tilde{s})\exp[\text{Pe}_1(\tfrac{1}{2} + p_1)\chi], \qquad \chi \leq 0 \tag{5.16}$$

$$\bar{\mathcal{R}}_3 = A_3(\tilde{s})\exp[\text{Pe}_3(\tfrac{1}{2} - p_3)\chi], \qquad \chi \geq \chi_1 \tag{5.17}$$

where

$$p_n = [\tfrac{1}{4} + (\tilde{s}/\text{Pe}_n)]^{1/2}, \qquad n = 1, 3 \tag{5.18}$$

$\tilde{s} = st_0$, the dimensionless variable in the Laplace transformation

$$\bar{f}(\tilde{s}) = \int_0^\infty [\exp(-\tilde{s}\theta)]f(\theta)\,d\theta$$

and $A_n(\tilde{s})$, $n = 1, 3$, are constants to be determined.

On the other hand, the solution of Equation (5.8) is

$$\begin{aligned}\bar{\mathcal{R}}_2 = {} & A_2(\tilde{s})\exp[\text{Pe}_2(\tfrac{1}{2} - p_2)] + B_2(\tilde{s})\exp[\text{Pe}_2(\tfrac{1}{2}) + p_2)] \\ & - \{H(\chi - \chi_0)\exp[\text{Pe}_2(\chi - \chi_0)/2]\sinh[\text{Pe}_2\,p_2(\chi - \chi_0)]\}/p_2\end{aligned} \tag{5.19}$$

where $H(\chi - \chi_0)$ is the Heaviside unit function in space at $\chi = \chi_0$ and $A_2(\tilde{s})$ and $B_2(\tilde{s})$ are constants, and $0 < \chi < \chi_1$.

Straightforward algebraic treatment of these equations and conditions gives

$$A_1 = \frac{e^{-(1/2)\text{Pe}_2\chi_0}[(p_3 + p_2)e^{\text{Pe}_2 p_2(\chi_1 - \chi_0)} + (p_2 - p_3)e^{-\text{Pe}_2 p_2(\chi_1 - \chi_0)}]}{(p_2 + p_1)(p_3 + p_2)e^{\text{Pe}_2 p_2 \chi_1} - (p_2 - p_1)(p_2 - p_3)e^{-\text{Pe}_2 p_2 \chi_1}} \tag{5.20}$$

$$A_2 = A_1(p_2 - p_1)/2p_2 \tag{5.21}$$

$$B_2 = A_1(p_2 + p_1)/2p_2 \tag{5.22}$$

These can now be used to give expressions for $\bar{\mathcal{R}}_1$, $\bar{\mathcal{R}}_2$, and $\bar{\mathcal{R}}_3$ if desired. In particular, they allow the calculation of expressions for the zeroth, first, second, . . . moments of the distribution curve in the appropriate region:

when $\tilde{s} = 0$ [and so, by Equation (5.18), $p_n = \frac{1}{2}$] the algebra will simplify appreciably. The outcome is as follows.

(a) Zeroth Moment

The zeroth moment gives the fraction of the whole solute injected that passes (in infinite time) any particular position. From Equations (5.20)–(5.22), respectively, there arises

$$(A_1)_{s\to 0} \longrightarrow \exp(-\mathrm{Pe}_2\chi_0)$$

$$(A_2)_{s\to 0} \longrightarrow 0$$

$$(B_2)_{s\to 0} \longrightarrow \exp(-\mathrm{Pe}_2\chi_0)$$

Hence, from Equation (5.16) it follows that for region 1

$$(\bar{\mathfrak{R}}_1)_{s\to 0} \equiv (m_0)_1 = \exp[-(\mathrm{Pe}_2\chi_0 - \mathrm{Pe}_1\chi)], \qquad \chi < 0$$

while from Equation (5.19) it follows that for region 2: (a) for $0 < \chi < \chi_0$ (where the Heaviside unit function $= 0$)

$$(\bar{\mathfrak{R}}_2)_{s\to 0} \equiv (m_0)_2 = \exp[-\mathrm{Pe}_2(\chi_0 - \chi)]$$

while (b) for $\chi_0 < \chi < \chi_1$ (where the Heaviside unit function $= 1$)

$$(\bar{\mathfrak{R}}_2)_{s\to 0} \equiv (m_0)_2 = 1$$

For region 3 an expression for A_3 is not needed; it is numerically equal to the value of $\bar{\mathfrak{R}}_3(\chi_1{}^+) = \bar{\mathfrak{R}}_2(\chi_1{}^-)$, by Equation (5.15). Hence,

$$\begin{aligned}[\bar{\mathfrak{R}}_3(\chi)]_{s\to 0} &\equiv (m_0)_3 \\ &= [\bar{\mathfrak{R}}_2(\chi_1{}^-)]_{s\to 0}\{\exp[\mathrm{Pe}_3(\tfrac{1}{2} - p_3)\chi]\}_{s\to 0}\end{aligned}$$

by Equation (5.17)

$$[\bar{\mathfrak{R}}_3(\chi)]_{s\to 0} = 1 \times 1$$

That is to say, all the injected solute will eventually pass all points beyond the injection point χ_0.

(b) Higher Moments

The first, second, . . . differential coefficients of $\bar{\mathfrak{R}}_n$ with respect to s can now be found and taken to the limit $\tilde{s} = 0$. If region 2 is considered, then Equation (5.19) combined with (5.20)–(5.22) can be so treated to yield expressions for $[\bar{\mathfrak{R}}_2{}'(\chi)]_{s=0}$, $[\bar{\mathfrak{R}}_2''(\chi)]_{s=0}$, . . . , i.e., the first, second, . . . , moments at any point downstream of the injection point. Now, these moments, as has already been said [see Equations (4.3) and (4.4)], are related to the differences of mean time of passage and to the differences of the variance of the concentration wave, and these can be computed from the shape of the concentration curve. Since for the present case the delta function, at its instant of irruption, is the zero of time and has zero variance and since $m_0 = 1$ [from Equation (4.2) *et seq.*] it follows that, at $\chi_m < \chi_0$,

$$m_1 = [\bar{\mathfrak{R}}_2'(\chi_m)]_{s=0} = \bar{\theta}$$

$$m_2 - m_1^2 = m_2 - \bar{\theta}^2 \equiv [\bar{\mathfrak{R}}_2''(\chi_m)]_{s=0} = \sigma_\theta^2$$

the variance being in dimensions of (dimensionless time)^2 and the numerical and dimensional relation between this and the measured variance σ_t^2 [which will be in dimensions of (time)^2] is $\sigma_\theta^2 = \sigma_t^2/t_0^2$. The form of the expressions for $[\bar{\mathfrak{R}}_2'(\chi_m)]_{s=0}$ and $[\bar{\mathfrak{R}}_2''(\chi_m)]_{s=0}$ depends upon the conditions, i.e., upon the values of Pe_1 and Pe_3 in relation to Pe_2; for a number of cases van der Laan [7] presents the expressions that give $\bar{\theta}$ and σ_θ^2, and these are reproduced in Table 5.1 (also see Figure 5.3). Levenspiel and Bischoff [8] also give a summary. In the general case considered here the expressions for $\bar{\theta}$ and σ_θ^2 are long and complicated and the finding of a value of Pe_2 (for that is what is usually sought) is correspondingly difficult. Other cases are less complicated (if, for example, $D_1 = D_3 = 0$ or D_2), but they may be difficult to perform experimentally. As an illustration, if $D_1 = D_2 = D_3$, the expressions are much simplified; this is the case treated by Levenspiel and Smith [9] and van der Laan reworked their example, using the present method. This example is treated in two ways: (i) on the assumption that L/U ($= t_0$) is known; (ii) that L/U is not known.

(i) *If t_0 is known*, then, since $\bar{t}$ ($= \int_0^\infty vt\,dt/\int_0^\infty v\,dt$) can be found from the measured values of v (in any units) against t (in time units), it follows that $\bar{\theta}$ ($= \bar{t}/t_0$) can be calculated. It is stressed that $t_0 = \bar{t}$ only for special condi ditions: this can be seen from Table 5.1, which gives some relations between $\bar{\theta}$ and Pe which may allow Pe to be found. One needs to get dimensionless mean time and, possibly, variance from the dimensional quantities, both of which can be computed from the concentration–time curve.

Numerical Example

Levenspiel and Smith found t_0 to be 7.7 sec, and $\bar{t}$ was 11.02 sec. Hence, $\bar{\theta} = 11.02/7.7 = 1.432 = (1 + 2/\text{Pe})$ (by Case 7 of Table 5.1). Hence, $1/\text{Pe} = 0.216$ (Levenspiel and Smith found $1/\text{Pe} = 0.217$.)

(ii) *If t_0 is not known*, then the above simple procedure for getting the dimensionless quantities is no longer possible. However, the ratio of the square of the mean time to the variance is the same for both dimensional and dimensionless quantities. Now, Levenspiel and Smith found that $\sigma_t^2 = 48.07$ sec^2. Hence,

$$\frac{\bar{\theta}^2}{\sigma_\theta^2} \equiv \frac{\bar{t}^2}{\sigma_t^2} = \frac{(11.02)^2}{(48.07)^2} = \frac{(1 + 2/\text{Pe})^2}{(2\text{Pe} + 8)/\text{Pe}^2}$$

by Case 7 in Table 5.1. Hence $1/\text{Pe}$ is computed from this to be 0.218.

TABLE 5.1

MEAN TIMES AND VARIANCES OF A DEGENERATE DIRAC IMPULSE: VARIOUS BOUNDARY CONDITIONS[a]

Case[b]	$\bar{\theta}$	σ_θ^2
1	$1 + (1/\mathrm{Pe})[2 - (1-\alpha)e^{-\mathrm{Pe}\chi_0} - (1-\beta)e^{-\mathrm{Pe}(\chi_1-\chi_m)}]$	$(1/\mathrm{Pe}^2)\{2\mathrm{Pe} + 8 + 2(1-\alpha)(1-\beta)e^{-\mathrm{Pe}\chi_1} - (1-\alpha)e^{-\mathrm{Pe}\chi_0}[4\chi_0\mathrm{Pe} + 4(1+\alpha) + (1-\alpha)e^{-\mathrm{Pe}\chi_0}] - (1-\beta)e^{-\mathrm{Pe}(\chi_1-\chi_m)}[4(\chi_1-\chi_m)\mathrm{Pe} + 4(1+\beta) + (1-\beta)e^{-\mathrm{Pe}(\chi_1-\chi_m)}]\}$
2	$1 + (1/\mathrm{Pe})[2 - e^{-\mathrm{Pe}\chi_0} - e^{-\mathrm{Pe}(\chi_1-\chi_m)}]$	$(1/\mathrm{Pe}^2)\{2\mathrm{Pe} + 8 + 2e^{-\mathrm{Pe}\chi_1} - e^{-\mathrm{Pe}\chi_0}(4\chi_0\mathrm{Pe} + 4 + e^{-\mathrm{Pe}\chi_0}) - e^{-\mathrm{Pe}(\chi_1-\chi_m)}[4(\chi_1-\chi_m)\mathrm{Pe} + 4 + e^{-\mathrm{Pe}(\chi_1-\chi_m)}]\}$
3	$1 + (1/\mathrm{Pe})[\alpha + \beta]$	$(1/\mathrm{Pe}^2)[2\mathrm{Pe} - 2 + 2e^{-\mathrm{Pe}} + 2(\alpha+\beta)(1-e^{-\mathrm{Pe}}) + 3(\alpha^2+\beta^2) + 2\alpha\beta e^{-\mathrm{Pe}}]$
4	1	$(2/\mathrm{Pe}^2)[\mathrm{Pe} - 1 + e^{-\mathrm{Pe}}]$
5	$1 + (1/\mathrm{Pe})[2 - (1-\alpha)e^{-\mathrm{Pe}\chi_0}]$	$(1/\mathrm{Pe}^2)\{2\mathrm{Pe} + 8 - (1-\alpha)e^{-\mathrm{Pe}\chi_0}[4\chi_0\mathrm{Pe} + 4(1+\alpha) + (1-\alpha)e^{-\mathrm{Pe}\chi_0}]\}$
6	$1 + (1/\mathrm{Pe})[2 - (1-\beta)e^{\mathrm{Pe}(\chi_1-\chi_m)}]$	$(1/\mathrm{Pe}^2)\{2\mathrm{Pe} + 8 - (1-\beta)e^{-\mathrm{Pe}(\chi_1-\chi_m)}[4(\chi_1-\chi_m)\mathrm{Pe} + 4(1+\beta) + (1-\alpha)e^{-\mathrm{Pe}(\chi_1-\chi_m)}]\}$
7	$1 + (2/\mathrm{Pe})$	$(1/\mathrm{Pe}^2)[2\mathrm{Pe} + 8]$
8	$1 + (1/\mathrm{Pe})$	$(1/\mathrm{Pe}^2)[2\mathrm{Pe} + 3]$

[a]Reprinted with permission of E. T. van der Laan [7], p. 189. Copyright 1958 by Pergamon Press.
[b]See Figure 5.3 for key.

Case	(Flow is left to right.)
1	0 X_0 X_m X_1 / D_a D D_b
2	0 X_0 X_m X_1 / $D_a = 0$ D $D_b = 0$
3	$X_0 = 0$ $X_m = X_1$ / D_a D D_b
4	$X_0 = 0$ $X_m = X_1$ / $D_a = 0$ D $D_b = 0$
5	0 X_0 X_m / D_a D
6	X_0 X_m X_1 / D D_b
7	X_0 X_m / D
8	$X_0 = 0$ X_m / $D_a = 0$ D / X_0 $X_m = X_1$ / D $D_b = 0$

Fig. 5.3. Key for Table 5.1.

Additionally (for the same case), $t_0 \equiv \bar{t}/\bar{\theta} = \bar{t}/(1 + 2/\text{Pe}) = 11.02/(1 + 2/\text{Pe})$. From this, t_0 is computed to be 7.67.

Both the values of $1/\text{Pe}$ and of t_0 given in the last paragraph can be compared with the values calculated by Levenspiel and Smith, which are given above (in the paragraph exemplifying the case when t_0 is known).

5.2.2 Generalized Impulse: One or Two Boundaries

The practical difficulty of utilizing the above treatment is that of generating a true impulse—or even a close approximation to it. Aris [10] pointed out that it was unnecessary to essay this provided that two measuring stations were set up downstream of a source of any shaped impulse (the shape being unknown). His reasoning, with the algebra corrected by Bischoff [11], is presented here. In Figure 5.4 the two measuring stations are at z_0 and z_m, both upstream of the boundary at z_1 and with no discontinuity or other change of dispersion coefficient between them. (On the other hand, when a perfect impulse was being considered, z_0 denoted the point of injection.) Any discontinuity that lies upstream of z_0 is irrelevant because the measured concentration–time curve at z_0 is used as a basis of comparison.

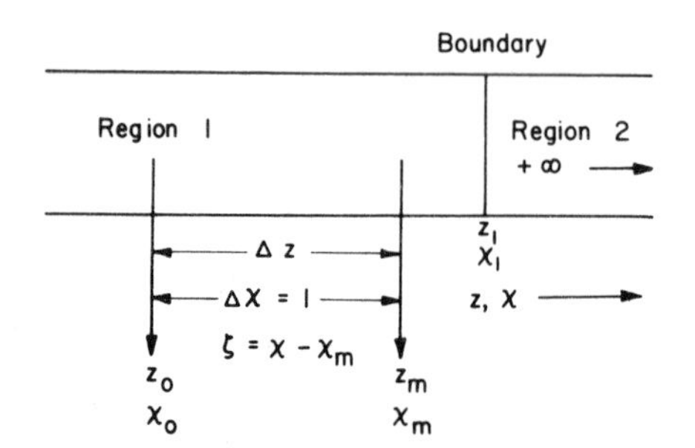

Fig. 5.4. One longitudinal boundary; moments of a generalized pulse at two observation points.

For $\chi_0 < \chi \leq \chi_1$

$$\left(\frac{1}{\mathrm{Pe}_1}\right)\frac{\partial^2 v_1}{\partial \chi^2} - \frac{\partial v_1}{\partial \chi} - \frac{\partial v_1}{\partial \theta} = 0 \tag{5.23}$$

and for $\chi > \chi_1$

$$\left(\frac{1}{\mathrm{Pe}_2}\right)\frac{\partial^2 v_2}{\partial \chi^2} - \frac{\partial v_2}{\partial \chi} - \frac{\partial v_2}{\partial \theta} = 0 \tag{5.24}$$

The conditions are as Equations (5.9), (5.13), and (5.15) with, in this instance, v_1 substituted for $\mathcal{R}_2$, v_2 substituted for $\mathcal{R}_3$, and with $\mathcal{R}_1$ omitted, while $L = z_m - z_0$ as in Figure 5.2, but z_0 is now the location of the first measuring point.

The Laplace transforms of Equations (5.23) and (5.24) give

$$\bar{v}_1 = A_1(\tilde{s}) \exp[\mathrm{Pe}_r(\tfrac{1}{2} + p_r)\chi] + B_1(\tilde{s}) \exp[\mathrm{Pe}_r(\tfrac{1}{2} - p_r)\chi], \quad r = 1, 2 \tag{5.25}$$

in which $p = [\frac{1}{4} + (\tilde{s}/\mathrm{Pe})]^{1/2}$ and $\tilde{s} = st_0$ (dimensionless). The condition at infinity gives $A_2 = 0$, and so

$$\bar{v}_2 = B_2 \exp[\mathrm{Pe}_r(\tfrac{1}{2} - p_r)\chi] \tag{5.26}$$

while the two conditions at χ_1 give

$$\frac{\partial \bar{v}_1(\chi_1{}^-)}{\partial \chi} = \frac{d\bar{v}_2(\chi_1{}^+)}{\partial \chi}\frac{\mathrm{Pe}_1}{\mathrm{Pe}_2}$$

which, applied to Equations (5.25) and (5.26), gives

$$[A_1(\tfrac{1}{2} + p_1)] \exp[\mathrm{Pe}_1(\tfrac{1}{2} + p_1)\chi_1] + B_1(\tfrac{1}{2} - p_1) \exp[\mathrm{Pe}_1(\tfrac{1}{2} - p_1)\chi_1] \\ - B_2(\tfrac{1}{2} - p_2) \exp[\mathrm{Pe}_2(\tfrac{1}{2} - p_2)\chi_1] = 0 \tag{5.27}$$

and

$$A_1 \exp[\mathrm{Pe}_1(\tfrac{1}{2} + p_1)\chi_1] + B_1 \exp[\mathrm{Pe}_1(\tfrac{1}{2} - p_1)\chi_1] \\ - B_2 \exp[\mathrm{Pe}_2(\tfrac{1}{2} - p_2)\chi_1] = 0 \tag{5.28}$$

while at χ_0,

$$A_1 \exp[\mathrm{Pe}_1(\tfrac{1}{2} + p_1)\chi_0] + B_1 \exp[\mathrm{Pe}_1(\tfrac{1}{2} - p_1)\chi_0] = \bar{v}_0 \tag{5.29}$$

where $\bar{v}_0 = \bar{v}(\chi_0, s)$ denotes the Laplace transform of the concentration, as a function of time, measured at χ_0. These three linear, nonhomogeneous equations allow the three unknown constants A_1, B_1, and B_2 to be found (in terms of $\bar{v}_0$) as

$$A_1 = \bar{v}_0(s) \frac{(p_1 - p_2) \exp[-\mathrm{Pe}_1(\frac{1}{2} + p_1)\chi_1]}{\text{Denominator}} \tag{5.30}$$

$$B_1 = \bar{v}_0(s) \frac{(p_1 + p_2) \exp[-\mathrm{Pe}_1(\frac{1}{2} - p_1)\chi_1]}{\text{Denominator}} \tag{5.31}$$

$$B_2 = \bar{v}_0(s) \frac{2p_1 \exp[-\mathrm{Pe}_2(\frac{1}{2} - p_1)\chi_1]}{\text{Denominator}} \tag{5.32}$$

where Denominator means

$$(p_1 - p_2) \exp[-\mathrm{Pe}_1(\tfrac{1}{2} + p_1)(\chi_1 - \chi_0)] \\ + (p_1 + p_2) \exp[-\mathrm{Pe}_1(\tfrac{1}{2} - p_1)(\chi_1 - \chi_0)]$$

Thus, in particular, at $\chi = \chi_\mathrm{m}$ ($\chi_0 \leq \chi_\mathrm{m} \leq \chi_1$) substitution for A_1 and B_1 into Equation (5.25) gives

$$\bar{v}_\mathrm{m}(s) = \bar{v}_0(s) \frac{(p_1 - p_2) \exp[-\mathrm{Pe}_1(\frac{1}{2} + p_1)(\chi_1 - \chi_\mathrm{m})] + (p_1 + p_2) \exp[-\mathrm{Pe}_1(\frac{1}{2} - p_1)(\chi_1 - \chi_\mathrm{m})]}{(p_1 - p_2) \exp[-\mathrm{Pe}_1(\frac{1}{2} + p_1)(\chi_1 - \chi_0)] + (p_1 + p_2) \exp[-\mathrm{Pe}_1(\frac{1}{2} - p_1)(\chi_1 - \chi_0)]} \tag{5.33}$$

i.e.,

$$\bar{v}_\mathrm{m}(s) = [\bar{v}_0(s)] \times [F(\chi_1, \chi_\mathrm{m}, \bar{s})] \tag{5.34}$$

The further development of the analysis requires the expansion of Equation (5.34) by Maclaurin's theorem. This states that a function $f(x)$ may be expanded as

$$f(x) = f(0) + xf'(0) + (x^2/2!)f''(0) + \cdots + (x^n/n!)f^{(n)}(0) + \cdots$$

the differentiation being with respect to x, and $f^{(n)}(0)$ meaning the value of the function obtained by putting $x = 0$ after differentiating the function n times, $n = 0, 1, 2, \ldots$. Hence, by this expansion,

$$\bar{v}(s) = \bar{v}(0) + s\bar{v}'(0) + (s^2/2!)\bar{v}''(0) + \cdots + (s^n/n!)\bar{v}^{(n)}(0) + \cdots$$

But Equation (4.8) stated that $\bar{v}^{(n)}(0) = (-1)^n m_n^*$ in general and $\bar{v}(0) = m_0$ in particular, m_n^* being the nth moment of the concentration distribution about the origin. Hence,

$$\begin{aligned} \bar{v}(s)/m_0 \equiv \bar{v}(s)/\bar{v}(0) &= 1 - (m_1^*/m_0)s + (m_2^*/m_0)(s^2/2) \\ &\quad - (m_3^*/m_0)(s^3/3!) + \cdots + (-t)^n(m_n^*/m_0)(s^n/n!) \\ &= 1 - \bar{t}s + (\sigma^2 + \bar{t}^2)(s^2/2!) - \cdots \end{aligned}$$

[by Equations (4.3) and (4.4)].

In particular, the component parts of Equation (5.34) can be written as

$$\bar{v}_{m}(s) = (m_0)_m[1 - \bar{t}_m s + (\sigma_m{}^2 + \bar{t}_m{}^2)(s^2/2!) - \cdots] \quad (5.35)$$

$$\bar{v}_0(s) = (m_0)_0[1 - \bar{t}_0 s + (\sigma_0{}^2 + \bar{t}_0{}^2(s^2/2!) - \cdots] \quad (5.36)$$

$$F(\bar{s}) = F(0) - \bar{t}_F s + (\sigma_F{}^2 + \bar{t}_F{}^2)(s^2/2!) - \cdots \quad (5.37)$$

in which $F(0)$, $\bar{t}_F$, and $\sigma_F{}^2$ are known algebraic expressions, obtained from Equation (5.33), as $F(0) = 1$,

$$\bar{t}_F \equiv -(dF/ds)_{s\to 0} \quad (5.38)$$

and

$$\sigma_F{}^2 \equiv [(d^2F/ds^2)_{s\to 0} - (dF/ds)^2_{s\to 0}] \quad (5.39)$$

Now, $(m_0)_m = (m_0)_0$ because all points considered are downstream of the injection point—see Section 5.2.1b—and so

$$\begin{aligned}
&[1 - \bar{t}_m s + (\sigma_m{}^2 + \bar{t}_m{}^2)(s^2/2!) - \cdots] \\
&\quad = [1 - \bar{t}_0 s + (\sigma_0{}^2 + \bar{t}_0{}^2)(s^2/2!) - \cdots][1 - \bar{t}_F \bar{s} + (\sigma_F{}^2 \\
&\qquad + \bar{t}_F{}^2)(\bar{s}^2/2!) - \cdots] \\
&\quad = \{1 - (\bar{t}_0 + \bar{t}_F t_0)s + [\sigma_0{}^2 + \sigma_F{}^2 + (\bar{t}_F t_0 + \bar{t}_0)^2](s^2/2!) - \cdots\}
\end{aligned}$$

The equating of like powers of s then gives

$$\bar{t}_F t_0 = \bar{t}_m - \bar{t}_0 \quad (5.40)$$

and

$$\sigma_F{}^2 t_0{}^2 = \sigma_m{}^2 - \sigma_0{}^2 + \bar{t}_m{}^2 - (\bar{t}_F t_0 + \bar{t}_0)^2$$

i.e.,

$$\sigma_F{}^2 t_0{}^2 = \sigma_m{}^2 - \sigma_0{}^2 \quad (5.41)$$

Thus, the differences between means and variances measured at the two stations give the numerical values of the quantities $\bar{t}_F t_0$ and $\sigma_F{}^2 t_0{}^2$.* But algebraic expressions for these two quantities can be obtained by virtue of Equations (5.38) and (5.39). These differentiations give, after some algebra, the following results for the cases shown in Figure 5.5:

Case (a)

$$\bar{t}_F = 1 - \{(1 - \beta)[1 - \exp(-\mathrm{Pe})] \exp(-\mathrm{Pe}\zeta)\}/\mathrm{Pe} \quad (5.42)$$

*In dimensions of (time) and (time)^2 respectively, when $\bar{t}$ and σ^2 are computed in the same dimensions. If mean time and variance are computed in (dimensionless time) then $\bar{t}_F$ $(= \bar{t}_m/t_0 - \bar{t}_0/t_0 = \bar{\theta}_m - \bar{\theta}_0)$ and $\sigma_F{}^2$ $[= \sigma_m{}^2/t_0{}^2 - \sigma_0{}^2/t_0{}^2 = (\sigma_\theta{}^2)_m - (\sigma_\theta{}^2)_0]$ are in (dimensionless time).

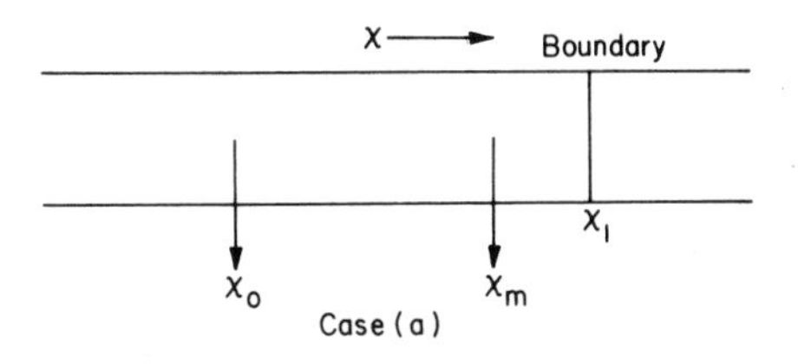

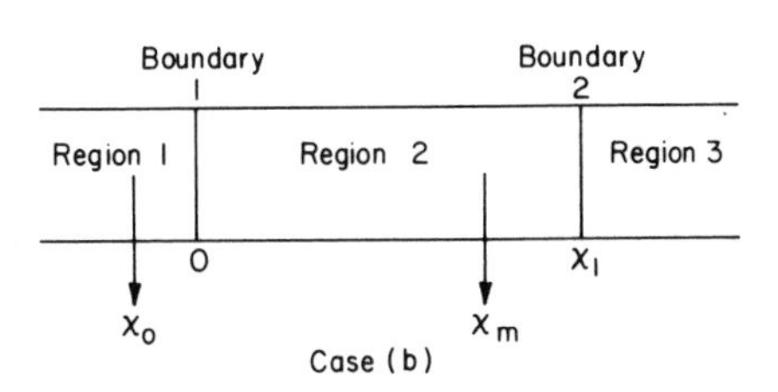

Fig. 5.5. Examples of systems subjected to an undefined pulse.

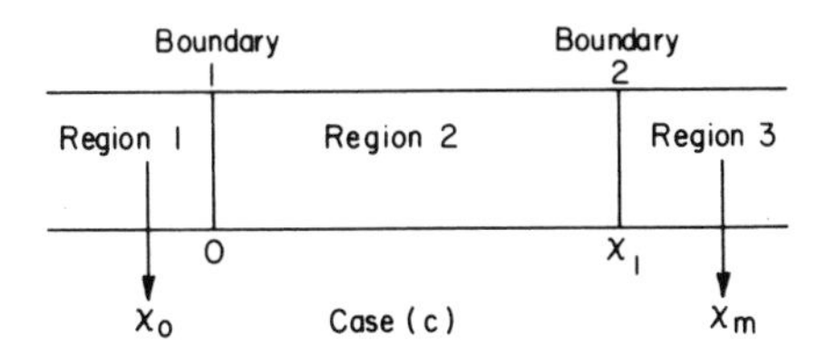

and

$$\begin{aligned}\sigma_F{}^2 &= (2/\text{Pe}) + [(1-\beta)/\text{Pe}^2]\{(1-\beta)[\exp(-2\text{Pe}\zeta)][\exp(-2\text{Pe})-1] \\ &\quad + 4(1+\beta)[\exp(-\text{Pe}\zeta)][\exp(-\text{Pe})-1] + 4\text{Pe}[\exp(-\text{Pe}\zeta)] \\ &\quad \times [(\zeta+1)\exp(-\text{Pe})-\zeta]\}\end{aligned} \tag{5.43}$$

Case (*b*) (Bischoff and Levenspiel [12])

$$\begin{aligned}\bar{t}_F &= 1 + [(1-\alpha)/\text{Pe}][1-\exp(\text{Pe}_1\chi_0)] + [(1-\beta)/\text{Pe}]\{\exp[\text{Pe}(\chi_m-\chi_1)]\} \\ &\quad \times [\exp(\text{Pe}_1\chi_0 - \text{Pe}\chi_m) - 1]\end{aligned} \tag{5.44}$$

and

$$\begin{aligned}\sigma_F{}^2 &= (2/\text{Pe}) + [1/\text{Pe}^2]\{3 + 2\alpha - 5\alpha^2 + 2(1-\alpha)\text{Pe}\chi_0 + 4(1-\alpha)\text{Pe}\chi_0 \\ &\quad \times [\exp(\text{Pe}_1\chi_0)] - 4(1-\alpha^2)[\exp(\text{Pe}_1\chi_0)] + 4(1-\beta^2)[\exp(-\text{Pe}\chi_1)] \\ &\quad \times [\exp(\text{Pe}_1\chi_0) - \exp(\text{Pe}\chi_m)] + 2(1-\alpha)(1-\beta)[\exp(-\text{Pe}\chi_1)] \\ &\quad \times [1-\exp(2\text{Pe}_1\chi_0)] + (1-\alpha)^2[\exp(2\text{Pe}_1\chi_0)] + 4(1-\beta) \\ &\quad \times [\exp(-\text{Pe}\chi_1)][\text{Pe}(\chi_m-\chi_1)\exp(\text{Pe}\chi_m)] - [\text{Pe}(\chi_0-\chi_1)\exp(\text{Pe}_1\chi_0)] \\ &\quad + (1-\beta)^2[\exp(-2\text{Pe}\chi_1)][\exp(2\text{Pe}_1\chi_0) - \exp(2\text{Pe}\chi_m)]\}\end{aligned} \tag{5.45}$$

Case (*c*) (Bischoff and Levenspiel [12])

$$\bar{t}_F = 1 + [(1 - \alpha)/\text{Pe}][1 - \exp(\text{Pe}_1\chi_0)] - [(1 - \beta)/\text{Pe}] \times [1 - \exp(\text{Pe}_1\chi_0 - \text{Pe}\chi_1)] \quad (5.46)$$

and

$$\begin{aligned}\sigma_F{}^2 = {} & (2/\text{Pe}) + (1/\text{Pe}^2)\{-2 + 2\alpha - 5\alpha^2 + 2\beta + 3\beta^2 + 2(1 - \beta) \\ & \times \text{Pe}(\chi_1 - \chi_m) + 2(1 - \alpha)\text{Pe}\chi_0 + 4(1 - \alpha)(\text{Pe}\chi_0 - 1 - \alpha) \\ & \times [\exp(\text{Pe}_1\chi_0)] + 4(1 - \beta)[1 + \beta - \text{Pe}(\chi_0 - \chi_1)][\exp(\text{Pe}_1\chi_0 \\ & - \text{Pe}\chi_1) + (1 - \beta)^2[\exp 2(\text{Pe}_1\chi_0 - \text{Pe}\chi_1)] + (1 - \alpha)^2[\exp(2\text{Pe}_1\chi_0)] \\ & + 2(1 - \alpha)(1 - \beta)[\exp(-\text{Pe}\chi_1)][1 - \exp(2\text{Pe}_1\chi_0)]\} \end{aligned} \quad (5.47)$$

In Equations (5.44)–(5.47) inclusive the symbol Pe is written for Pe_2 (the value in the central region, taken to be the one of interest) for convenience. Thus, by using Equations (5.42), (5.44), and (5.46), the value of Pe_2 in the respective case can be found if $t_0 = U/L$ is known and provided also that α and β are known.

If, however, t_0 is not known, then both the expressions for $\bar{t}_F$ and $\sigma_F{}^2$ must be used. These give rise to two equations in the two unknowns Pe_2 and t_0, and graphic or other methods should allow both to be found.

5.3 THE GENERAL CASE: GENERALIZED IMPULSE OR INITIAL DISTURBANCE, *n* BOUNDARIES

The situations discussed in Sections 5.2.1 and 5.2.2, seemingly different, can in fact be seen to be particular examples of a generalization outlined in the present section.

Consider a one-dimensional flow system having $(n + 1)$ *regions* between *n boundaries*, as shown in Figure 5.6, the first and last sections obviously extending to $-\infty$ and $+\infty$, respectively (otherwise there would be boundaries beyond the "first" and "last").

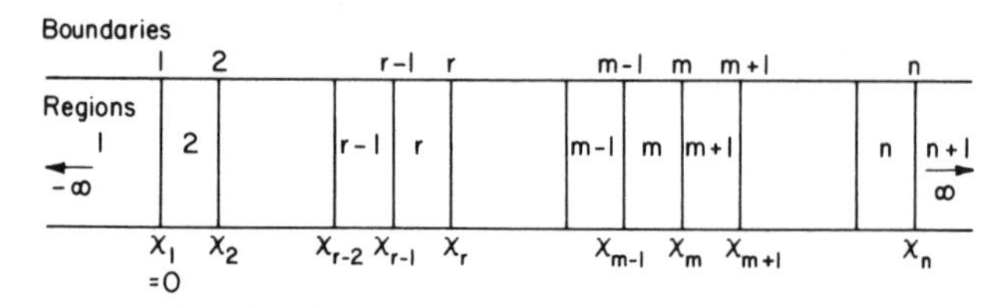

Fig. 5.6. General case; *n* regions, undefined input.

In any section a differential balance is

$$\frac{\partial^2 v}{\partial \chi^2} - \mathrm{Pe}\frac{\partial v}{\partial \chi} - \mathrm{Pe}\frac{\partial v}{\partial \theta} = f(\chi, \theta) \tag{5.48}$$

$f(\chi, \theta)$ representing a generalized source term at χ. The Laplace transform of Equation (5.48) (the transform being $\bar{v} = \int_0^\infty [\exp(-\bar{s}\theta)]v(\theta)\, d\theta$, where $\bar{s} = st_0 = sL/U$, L being any reference length) gives

$$\frac{d^2\bar{v}}{d\chi^2} - \mathrm{Pe}\frac{d\bar{v}}{d\chi} - \bar{s}\ \mathrm{Pe}\ \bar{v} = \epsilon \tag{5.49}$$

where

$$\epsilon = \bar{f}(\bar{s}, \chi) - \mathrm{Pe}\ v(0, \chi) \tag{5.50}$$

and subject to the conditions discussed by Aris [13]. Now, for at least one region, ϵ must be nonzero, for if it were zero everywhere, then both $\bar{f}$ and $v(0, \chi)$ would be zero everywhere: if $\bar{f}$ were zero, then there can be no source term; while if $v(0, \chi) = 0$, there cannot be an initial nonuniform concentration distribution.

If $v(0, \chi)$ is not uniform in any one region, then it is not uniform in any other region, for by Wehner and Wilhelm's conditions (Section 5.1), there can be no concentration discontinuity at a boundary. So, for the rth region

$$\frac{d^2\bar{v}_r(\chi, \bar{s})}{d\chi^2} - \mathrm{Pe}_r\frac{d\bar{v}_r(\chi, \bar{s})}{d\chi} - \bar{s}\ \mathrm{Pe}_r\ \bar{v}_r = \epsilon_r(\chi, \bar{s}, \mathrm{Pe}_r) \tag{5.51}$$

where

$$\epsilon_r = \bar{f}(\chi, \bar{s}) - \mathrm{Pe}_m\ v_m(0, \chi) \qquad \text{for} \quad r = m \tag{5.52}$$

and

$$\epsilon_r = -\mathrm{Pe}_r\ v_r(0, \chi) \qquad \text{for} \quad r \neq m \tag{5.53}$$

The solution to Equation (5.51) is

$$\bar{v}_r = A_r(\exp \mu_r\chi) + B_r(\exp \nu_r\chi) + E_r(\chi, \bar{s}, \mathrm{Pe}_r) \tag{5.54}$$

where E_r is the Particular Integral of $\epsilon_r(\chi, \bar{s}, \mathrm{Pe}_r)$ with respect to χ,

$$\mu_r = \mathrm{Pe}_r(\tfrac{1}{2} + p_r) < 0$$
$$\nu_r = \mathrm{Pe}_r(\tfrac{1}{2} - p_r) > 0$$
$$p_r = [\tfrac{1}{4} + (\bar{s}/\mathrm{Pe}_r)]^{1/2}$$

The substitution of the Laplace transforms of the boundary conditions in Equations (5.10)–(5.15) into Equation (5.54) leads to

$$A_r(\exp \mu_r\chi_r) + B_r(\exp \nu_r\chi_r) + E_{r,r} = A_{r+1}(\exp \mu_{r+1}\chi_{r+1}) + B_{r+1}(\exp \nu_{r+1}\chi_r) + E_{r+1,r} \tag{5.55}$$

and

$$(\tfrac{1}{2} + p_r)A_r(\exp \mu_r\chi_r) + (\tfrac{1}{2} - p_r)B_r(\exp \nu_r\chi_r) + (1/\mathrm{Pe}_r)E'_{r,r}$$
$$= (\tfrac{1}{2} + p_{r+1})A_{r+1}(\exp \mu_{r+1}\chi_r) + (\tfrac{1}{2} - p_r)B_{r+1}(\exp \nu_{r+1}\chi_r)$$
$$+ (1/\mathrm{Pe}_{r+1})E'_{r+1,r} \tag{5.56}$$

where $r = 1, \ldots, n$ but $E_{m,m-1}$ and $E_{m,m}$ are different from the rest; [see Equations (5.52) and (5.53)]. In the above equations the first of the double suffixes refers to the region and the second refers to the boundary. For convenience new symbols are introduced, viz.,

$$\exp \mu_j\chi_r = a_{jr} \tag{5.57a}$$
$$(\tfrac{1}{2} + p_j) \exp \mu_j\chi_r = a'_{jr} \tag{5.57b}$$
$$\exp \nu_j\chi_r = b_{jr} \tag{5.57c}$$
$$(\tfrac{1}{2} - p_j) \exp \nu_j\chi_r = b'_{jr} \tag{5.57d}$$

where $j = r$ or $r + 1$; while

$$E_{r+1,r} - E_{r,r} = \psi_r \tag{5.57e}$$
$$(1/\mathrm{Pe}_{r+1})E'_{r+1,r} - (1/\mathrm{Pe}_r)E'_{r,r} = \psi_r', \qquad \text{for} \quad r = 1, \ldots, n \tag{5.57f}$$

but again ψ_{m-1} $(= E_{m,m-1} - E_{m-1,m-1})$ and ψ_m $(= E_{m-1,m} - E_{m,m})$ are different. Since $\mu < 0$, $\nu > 0$, it follows that $B_1 = A_{n+1} = 0$ while, if $\chi = 0$ at the first boundary, then $a_{11} = a_{21} = b_{11} = b_{21} = 1$, and $a'_{11} = \tfrac{1}{2} + p_1$, $a'_{21} = \tfrac{1}{2} + p_2$, $b'_{11} = \tfrac{1}{2} - p_1$, and $b'_{21} = \tfrac{1}{2} - p_2$.

The equations can now be written

$$A_1 - A_2 - B_2 = \psi_1 \tag{5.58(i)}$$
$$a'_{11}A_1 - a'_{21}A_2 - b'_{21}B_2 = \psi_1' \tag{5.58(ii)}$$
$$\vdots$$
$$a_{r,r}A_r + b_{r,r}B_r - a_{r+1,r}A_{r+1} - b_{r+1,r}B_{r+1} = \psi_r \tag{5.58($2r-1$)}$$
$$a'_{r,r}A_r + b'_{r,r}B_r - a'_{r+1,r}A_{r+1} - b'_{r+1,r}B_{r+1} = \psi_r' \tag{5.58($2r$)}$$
$$\vdots$$
$$a_{n,n}A_n + b_{n,n}B_n + b_{n+1,n}B_{n+1} = \psi_n \tag{5.58($2n-1$)}$$
$$a'_{n,n}A_n + b'_{n,n}B_n + b'_{n+1,n}B_{n+1} = \psi_n' \tag{5.58($2n$)}$$

Hence, from these $2n$ equations the values of the $2n$ "unknowns" A_r and B_r, $r = 1, \ldots, n$, can be found by setting up the determinants, using the coefficients a, a', b, b', ψ, ψ', in the usual way. The resulting algebraic expressions can then be put into Equation (5.54) to give equations for $\bar{v}_r$ that can be differentiated once, twice, . . . with respect to $\bar{s}$ as in Sections 5.2.1 and 5.2.2 in order to calculate the first, second, . . . moments of the distribution.

To proceed on these lines in the general case invokes much heavy algebra, particularly as the differential coefficients of the determinants have to be set out, A_r and B_r being functions of $\bar{s}$. One deduction can be made, however: it is that since A_r and B_r are functions of all parameters, then the distribution in each region depends on the A's and B's upstream and downstream of the region. Specific cases, either simple ones or with known numerical values, can be worked out, of course. For example, it can be seen that the cases discussed in Sections 5.2.1 and 5.2.2 are special cases of the above.

Thus, in Section 5.2.1 there are three sections, two boundaries, six unknown constants (viz., A_1, A_2, A_3 and B_1, B_2, B_3, four equations from the boundary conditions, two more from the conditions at $\pm\infty$. There was an impulsive source in region 2 at z_0, the initial distribution being uniform everywhere. Equation (5.19) is thus a specific example of Equation (5.54).

In Section 5.2.2 the treatment is generalized to the extent that the form of the source is not specified. In the first case treated, there are two regions with one boundary between them. The second region is of infinite length, but the first region is undefined. There are, accordingly, four unknown constants (A_1, A_2 and B_1, B_2), but only two equations from the one boundary, together with one more equation from the limiting condition at $z \longrightarrow +\infty$. The remaining equation, viz., (5.29), was obtained by writing

$$\bar{v}_0(\bar{s}) = A_1(\exp \mu_1 z_0) + B_1(\exp \nu_1 z_0)$$

where $\bar{v}_0(\bar{s})$ is the Laplace transform of $v_0(\theta)$, the measured distribution at $\chi = \chi_0$, the first measuring point. (Dimensionless time is used in the present section, in contrast to Section 5.2.2.) Hence, the equations are

$$a'_{11}A_1 - b'_{11}B_1 - b'_{21}B_2 = 0 \qquad (5.27)\ \mathrm{R}$$

$$a_{11}A_1 - b_{11}B_1 - b_{21}B_2 = 0 \qquad (5.28)\ \mathrm{R}$$

$$A_1(\exp \mu_1 z_0) + B_1(\exp \nu_1 z_0) = \bar{v}_0 \qquad (5.29)\ \mathrm{R}$$

and these enable expressions to be written down and solved for A_1, A_2, and B_2, all involving $\bar{v}_0$ and other parameters. Hence, the equation (again involving $\bar{v}_0$) for $\bar{v}(\bar{s})$ at the second measuring point can be written down. The subsequent argument—the finding of the first and second differential coefficients of $\bar{v}$ at the first and second measuring points—was explained in Section 5.2.2.

The remaining cases in Section 5.2.2 represent extensions of the last example, for here three sections and hence two boundaries between them require six constants and provide four; the condition at $\chi \longrightarrow +\infty$ and the incorporation of the distribution at the first measuring station again provide two more, The fact that there are three regions has made much lengthier the final expressions for $\sigma_F{}^2$ and $\bar{t}_F$.

5.4 THE GENERAL CASE: STEADY-STATE, FIRST-ORDER REACTION, *n* BOUNDARIES

It was mentioned in Section 5.1 that the use of Wehner and Wilhelm's boundary conditions in a reactor might be expected to lead to conclusions different from those drawn for a transient, nonreacting, system because the differential equations are, in general, different. Hence, it is of interest to examine a case analogous to the general transient case discussed in Section 5.3; that is to say, there are $n + 1$ sections, with n boundaries, the first and last extending to infinity. In the mth section a reaction is proceding; for simplicity in this illustration this will be assumed to be first-order and homogeneous, of rate constant K. The situation has the same numbering as in Figure 5.6.

In the following, χ is a dimensionless distance, equal to z/L, L being a reference length, conveniently that of the reacting section $\chi_m - \chi_{m-1}$; $\mathrm{Pe}_r = U_r L/D_r$, where U_r and D_r are velocities and longitudinal dispersion coefficients in the rth section; and $K = kL/U_m$, where k is the first-order reaction constant.

For all sections except the mth we have

$$(d^2 v_r/d\chi^2) - \mathrm{Pe}_r\, dv_r/d\chi = 0, \qquad r \neq m \tag{5.59}$$

with conditions at all boundaries

$$v_{r,r} = v_{r+1,r} \tag{5.60}$$

$$(1/\mathrm{Pe}_r)(dv_r/d\chi)_r - v_{rr} = (1/\mathrm{Pe}_{r+1})(dv_{r+1}/d\chi)_r - v_{r+1,r} \tag{5.61}$$

and the solution is

$$v_r = A_r + B_r \exp(\mathrm{Pe}_r\chi), \qquad r \neq m, \qquad \mathrm{Pe}_r \neq \infty \tag{5.62}$$

The differential equation for section m is

$$(d^2 v_m/d\chi^2) - \mathrm{Pe}_m(dv_m/d\chi) - \mathrm{Pe}_m K v_m = 0, \qquad r = m \tag{5.63}$$

with boundary conditions again given by Equations (5.60) and (5.61). The solution is

$$v_m = A_m(\exp \dot{\mu}_m\chi) + B_m(\exp \dot{\nu}_m\chi) \tag{5.64}$$

where

$$\dot{\mu}_m = \mathrm{Pe}_m(\tfrac{1}{2} + \dot{p}_m) \tag{5.65}$$

$$\dot{\nu}_m = \mathrm{Pe}_m(\tfrac{1}{2} - \dot{p}_m) \tag{5.66}$$

and

$$\dot{p}_m = [\tfrac{1}{4} + (K/\mathrm{Pe}_m)]^{1/2} \tag{5.67}$$

[cf. Equation (5.54); now, neither t nor s is involved].

The equations of the concentration in the rth and mth sections, viz., (5.62) and (5.64), combined with the boundary conditions given by Equations (5.60) and (5.61) lead to

$$A_r + B_r \exp(\mathrm{Pe}_r\chi_{r,r}) = A_{r+1} + B_{r+1}\exp(\mathrm{Pe}_{r+1}\chi_{r+1,r}) \tag{5.68}$$

$$B_r \exp(\mathrm{Pe}_r\chi_{r,r}) = B_{r+1}\exp(\mathrm{Pe}_{r+1}\chi_{r+1,r}), \qquad r \neq m \tag{5.69}$$

$$A_{m-1} + B_{m-1}\exp(\mathrm{Pe}_{m-1}\chi_{m-1,m-1}) = A_m \exp(\dot{\mu}_m\chi_{m,m-1}) + B_m \exp(\dot{\nu}_{m,}\chi_{m,m-1}) \tag{5.70}$$

$$B_{m-1}\exp(\mathrm{Pe}_{m-1}\chi_{m-1,m-1}) = (A_m\dot{\mu}_m/\mathrm{Pe}_m)\exp(\dot{\mu}_m\chi_{m,m-1}) + (B_m\dot{\nu}_m/\mathrm{Pe}_m)\exp(\dot{\nu}_m\chi_{m,m-1}) \tag{5.71}$$

$$A_m \exp(\dot{\mu}_m\chi_{m,m}) + B_m \exp(\dot{\nu}_m\chi_{m,m}) = A_{m+1} + B_{m+1}\exp(\mathrm{Pe}_{m+1}\chi_{m+1,m}) \tag{5.72}$$

$$(A_m\dot{\mu}_m/\mathrm{Pe}_m)\exp(\dot{\mu}_m\chi_{m,m}) + (B_m\dot{\nu}_m/\mathrm{Pe}_m)\exp(\dot{\nu}_m\chi_{m,m}) = B_{m+1}\exp(\mathrm{Pe}_{m+1}\chi_{m+1,m}) \qquad (5.73)$$

In the above equations the first (or only) suffix refers to a region, and the second suffix refers to a boundary so designated approached from the designated region.

Given also the stipulations that feed enters the system at $-\infty$ at concentration v_0, and that the concentration remains finite at A_b as $z \longrightarrow \infty$, then it may be deduced immediately from Equations (5.68) and (5.69) that

$$A_1 = A_2 = \cdots A_{m-1} \equiv v_0 \qquad (5.74)$$

$$A_{m+1} = A_{m+2} = \cdots A_{n+1} \equiv A_b \qquad (5.75)$$

and from Equation (5.62) that

$$B_{n+1} = 0 \qquad (5.76)$$

From Equations (5.76) and (5.69) it follows that $B_{n+1} = B_n = B_{n-1} = \cdots = B_{m+1} = 0$.

The situation now has resolved itself into a collection of $m + 2$ linear simultaneous equations in the $m + 2$ unknowns A_b, A_m, and $B_1, \ldots, B_m$. The coefficients of these unknowns are written, for convenience, as generally

$$b_{ir} = \exp \mathrm{Pe}_i\chi_r, \qquad i = r \text{ or } r+1, \quad r \neq m \qquad (5.77a)$$

while in the mth region

$$a_{mj} = \exp(\dot{\mu}_m\chi_j) = \exp[\mathrm{Pe}_m(\tfrac{1}{2} + \dot{p}_m)\chi_j] \qquad (5.77b)$$

$$\bar{a}_{mj} = (\dot{\mu}_m/\mathrm{Pe}_m)\exp(\dot{\mu}_m\chi_j) = (\tfrac{1}{2} + \dot{p}_m)\exp[(\tfrac{1}{2} + \dot{p}_m)\chi_j] \qquad (5.77c)$$

$$b_{mj} = \exp(\dot{\nu}_m\chi_j) = \exp[\mathrm{Pe}_m(\tfrac{1}{2} - \dot{p}_m)\chi_j] \qquad (5.77d)$$

$$\bar{b}_{mj} = (\dot{\nu}_m/\mathrm{Pe}_m)\exp(\dot{\nu}_m\chi_j) = (\tfrac{1}{2} - \dot{p}_m)\exp[(\tfrac{1}{2} - \dot{p}_m)\chi_j] \qquad (5.77e)$$

where $j \equiv m, m$ or $m, (m-1)$.

The equations given earlier now become, in terms of these new variables, those shown in Table 5.2.

TABLE 5.2

From Equation	Boundary number	Equation		Number
(5.69)	1	$b_{11}B_1 - b_{21}B_2$	$= 0$	(5.74(1))
			$\vdots$	
(5.69)	r	$b_{r,r}B - b_{r+1,r}B_{r+1}$	$= 0$	(5.74(r))
			$\vdots$	
(5.69)	$m-2$	$b_{m-2,m-2}B_{m-2} - b_{m-1,m-2}B_{m-1}$	$= 0$	(5.74($m-2$))
(5.70)	$m-1$	$b_{m-1,m-1}B_{m-1} - a_{m,m-1}A_m - b_{m,m-1}B_m$	$= -v_0$	(5.74($m-1$))
(5.71)		$b_{m-1,m-1}B_{m-1} - \bar{a}_{m,m-1}A_m - \bar{b}_{m,m-1}B_m$	$= 0$	(5.74(m))
(5.72)	m	$a_{m,m}A_m + b_{m,m}B_m - A_b$	$= 0$	(5.74($m+1$))
(5.73)		$\bar{a}_{m,m}A_m + \bar{b}_{m,m}B_m$	$= 0$	(5.74($m+2$))

A matrix of coefficients and constant terms is

$$\left|\begin{array}{ccccccc}
b_{11} - b_{21} & & & & & & \\
b_{22} & & & & & & \\
\cdots & & & 0 & & & \\
b_{r,r} - b_{r+1,r} & & & & & & \\
\cdots & & & & & & \\
- b_{m-1,m-2} & & & & & & \\
& b_{m-1,m-1} - b_{m,m-1} - a_{m,m-1} & & & 0 & -v_0 \\
0 & b_{m-1,m-1} - \bar{b}_{m,m-1} - \bar{a}_{m,m-1} & & & 0 & 0 \\
& b_{m,m} & a_{m,m} & & -1 & 0 \\
& \bar{b}_{m,m} & \bar{a}_{m,m} & & 0 & 0
\end{array}\right|$$

from which the unknowns $B_1, \ldots, B_m$, and A_m, A_b may be determined by Cramer's rule as the ratio of the two determinants d and D. The appropriate expressions are

$$D = -b_{11}b_{22} \cdots b_{m-1,m-1}\Delta_1$$

$$(d)_{A_m} = -b_{11}b_{22} \cdots b_{m-1,m-1}\bar{b}_{m,m}v_0$$

$$(d)_{B_m} = b_{11}b_{32} \cdots b_{m-1,m-1}\bar{a}_{m,m}v_0$$

$$(d)_{A_b} = b_{11}b_{22} \cdots b_{m-1,m-1}v_0\Delta_2$$

$$(d)_{B_r} = (-1)b_{11}b_{22} \cdots b_{r-1,r-1}v_0b_{r+1,r} \cdots b_{m-1,m-2}\Delta_3, \qquad r = 1, \cdots, m-1$$

where

$$\Delta_1 = \begin{vmatrix} (a_{m,m-1} - \bar{a}_{m,m-1}) & (b_{m,m-1} - \bar{b}_{m,m-1}) \\ \bar{a}_{m,m} & (\bar{b}_{m,m}) \end{vmatrix}$$

$$\Delta_2 = \begin{vmatrix} b_{m,m} & a_{m,m} \\ \bar{b}_{m,m} & \bar{a}_{m,m} \end{vmatrix}$$

$$\Delta_3 = \begin{vmatrix} \bar{b}_{m,m-1} & \bar{a}_{m,m-1} \\ \bar{b}_{m,m} & \bar{a}_{m,m} \end{vmatrix}$$

Hence,

$$A_m = \bar{b}_{m,m}v_0/\Delta_1$$

$$B_m = -\bar{a}_{m,m}v_0/\Delta_1$$

$$A_b = -v_0\Delta_2/\Delta_1$$

$$B_r = -v_0b_{r+1,r} \cdots b_{m-1,m-2}\Delta_3/b_{r,r} \cdots b_{m-1,m-1}\Delta_1$$

It follows from the above that the concentration increases upstream, and in each section it is governed by an exponential law. There is no discontinuity at any boundary, the limiting case of $\mathrm{Pe}_r \longrightarrow \infty$ tending to a distribution as in Figure 5.7. That is, it can be considered that there is no discontinuity in concentration across a boundary, but "infinite" slope; i.e., mathematically, "a discontinuity of the first kind." This was pointed out by Wehner and Wilhelm [5], whose example, in which $n + 1 = 3$, is a special case of the above, and can be so deduced. It will be noticed also that, no matter how many sections upstream or downstream of the reaction section, the concentration in the $m + 1, m + 2, \ldots, n + 1$ sections is constant with time, and that for all nonzero values of Pe_r, $r = m, m + 1, \ldots, n + 1$,

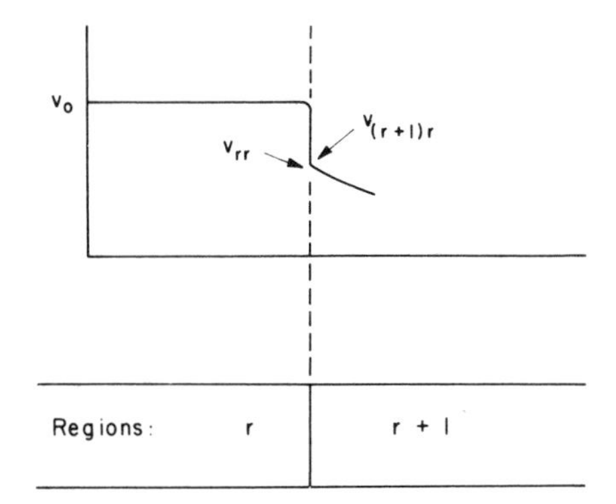

Fig. 5.7. Steady-state, first-order reaction; concentration across the inlet boundary.

the slope $dv_{m,m}/d\chi = 0$, as may be obtained from Equation (5.64). This is to be contrasted with the case of a transient state discussed earlier; even the envelope of a sinusoidal wave, itself constant with time, has a nonzero slope at all boundaries except at the rth boundary in the special case when $\mathrm{Pe}_{r+1} \longrightarrow \infty$, as will be discussed in Chapter 6.

REFERENCES

[1] H. M. Hulburt, Chemical Processes in Continuous Flow Systems. *Ind. Eng. Chem.* **36**, 1012 (1944).
[2] P. V. Danckwerts, Continuous Flow Systems. *Chem. Eng. Sci.* **2**, 1 (1953).
[3] L.-T. Fan and Y.-K. Ahn, Critical Evaluation of Boundary Conditions for Tubular Flow Reactors. *Ind. Eng. Chem. Proc. Design Develop.* **1**, 3 (1962).
[4] J. R. A. Pearson, A Note on the "Danckwerts" Boundary Conditions for Continuous Flow Reactors. *Chem. Eng. Sci.* **10**, 281 (1959).
[5] J. F. Wehner and R. H. Wilhelm, Boundary Conditions of Flow Reactor. *Chem. Eng. Sci.* **6**, 89 (1956).
[6] K. B. Bischoff, A Note on Boundary Conditions for Flow Reactors. *Chem. Eng. Sci.* **16**, 131 (1961).
[7] E. T. van der Laan, Notes on the Diffusion-Type Model for the Longitudinal Mixing in Flow. *Chem. Eng. Sci.* **7**, 187 (1958).
[8] O. Levenspiel and K. B. Bischoff, Patterns of Flow in Chemical Process Vessels. "Advances in Chemical Engineering" (T. B. Hoopes, J. W. Drew, and T. Vermeulen, eds.), Vol. 4. Academic Press, New York, 1963.
[9] O. Levenspiel and W. K. Smith, Notes on the Diffusion-type Model for the Longitudinal Mixing of Fluids in Flow. *Chem. Eng. Sci.* **6**, 277 (1957).
[10] R. Aris, Notes on the Diffusion-Type Model for Longitudinal Mixing in Flow. *Chem. Eng. Sci.* **9**, 266 (1959).
[11] K. B. Bischoff, Notes on the Diffusion-Type Model for Longitudinal Mixing in Flow. *Chem. Eng. Sci.* **12**, 69 (1960).
[12] K. B. Bischoff and O. Levenspiel, Fluid Dispersion-Generalization and Comparison of Mathematical Models: Part I. *Chem. Eng. Sci.* **17**, 245 (1962); Part II. *Ibid.* **17**, 257 (1962).
[13] R. Aris, On the Dispersion of Linear Kinematic Waves. *Proc. Roy. Soc.* (*London*) **A-245**, 268 (1958).

Chapter 6

Sinusoidal Waves

6.1 INTRODUCTION

In Chapter 1 were set out some wave equations. When the temperature or concentration varies sinusoidally the solutions to the wave equations [viz., (1.5a)–(1.5d)] were, respectively,

$$V = Ae^{\gamma_1 z} + Be^{\gamma_2 z} \tag{1.10a R}$$

if $D \neq 0$ and

$$V = Ae^{\gamma_1 z} \tag{1.10b R}$$

if $D = 0$. The point of note is that in a flow system γ_1 is not equal to γ_2.

The values of A and B can be found if conditions are specified that have to be satisfied; the ones to be discussed in this chapter are those shown in Figure

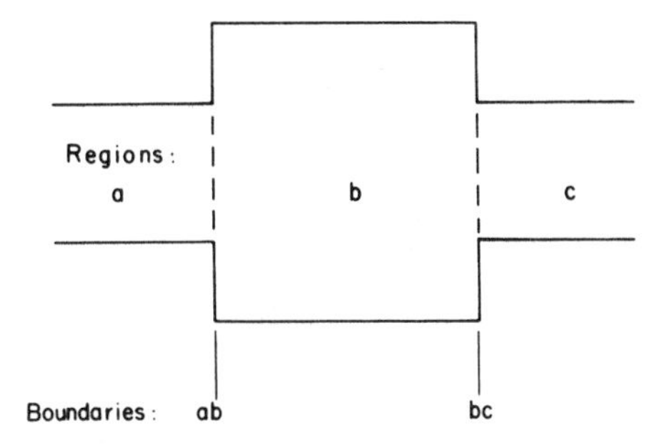

Fig. 6.1. Frequently occurring system: constants of integration have to satisfy conditions at two boundaries.

TABLE 6.1

SYMBOLS RELEVANT TO TRAVELING SINE WAVES[a]

$v = Ve^{i\omega t}, q = Qe^{i\omega t}$	Instantaneous value (resultant)
$v^+ = \mathfrak{v}^+e^{i\omega t}, q^+ = \mathfrak{q}^+e^{i\omega t}$	Instantaneous incident (positive) value
$v^- = \mathfrak{v}^-e^{i\omega t}, q^- = \mathfrak{q}^-e^{i\omega t}$	Instantaneous reflected (negative or echo) value
$\mathfrak{v}^+, \mathfrak{q}^+$	Incident (positive) phasor
$\mathfrak{v}^-, \mathfrak{q}^-$	Reflected (negative or echo) phasor
V^+, Q^+	Incident (positive) phasor at load (electrical) or exit boundary
V^-, Q^-	Reflected (negative or echo) phasor at load (electrical) or exit boundary
$V(z), Q(z)$	Resultant phasor at a general value of z
V_s, Q_s	Resultant phasor at sending end (electrical) or entrance boundary
V_R, Q_R	Resultant phasor at load (electrical) or exit boundary

[a]NOTES: (i) All v's may denote either voltage or a concentration of heat or matter. All q's may denote an electrical current or a flux of heat or matter per unit cross section. (ii) Positive waves travel in the direction of z increasing and vice versa.

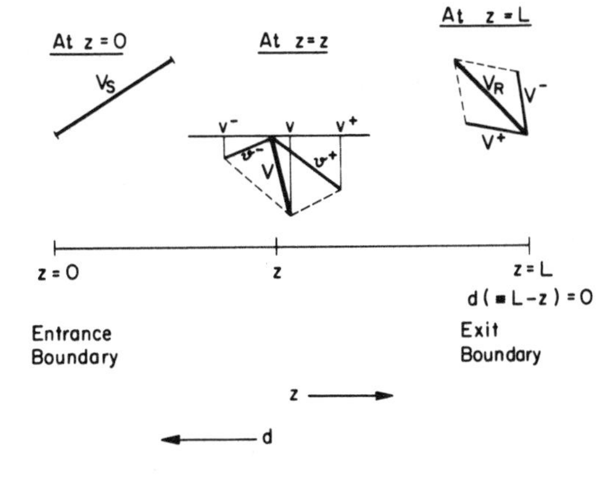

Fig. 6.2. Definition of symbols of concentration; instantaneous values and phasors. (Nomenclature for flux corresponds.) See Table 6.1.

6.1, being of very frequent occurrence. When they are known then the amplitude and phase angle of the wave at any point in the system can be found. However, neither of these quantities would be simple functions of distance, in contrast to the situation that would obtain if the system were infinitely long (discussed elsewhere); in that case both the amplitude and phase angle would be of a form that would make their use relatively straightforward. Thus, it would seem that the practical necessity of using a bed of finite length to make measurements would remove this most desirable feature.

It is possible, however, to overcome this complication by the relatively simple means of finding the frequency response of two lengths of the same system, all else being unchanged. The conditions are not at all severe; they and the justification for the method are given below. To arrive at them, it is convenient to introduce one or two concepts that are widely used in some branches of traveling wave engineering, viz., the telegrapher's equations, impedances, and reflections, as well as a collection of symbols. The latter are given in Table 6.1 and Figure 6.2.

6.2 THE TELEGRAPHER'S EQUATIONS

These date back to the burgeoning of the electrical telegraph when it was realized that fluctuating currents in a long line showed unexpected behavior. They relate a flux (current) to a driving force (voltage) by means of two differential equations. The *first telegrapher's equation* gives the longitudinal gradient of temperature or concentration (or voltage). It relates this to the flux of heat or mass (or electricity), and is an ordinary differential equation in z in the present case.

The *second telegrapher's equation* gives the longitudinal gradient of flux. It arises from a differential balance over length dz (viz., input = output plus internal increase) and so it is a partial differential equation in z and t.

Thus, from Figure 6.3(a) we have, when v denotes *concentration*,

$$q = -D(\partial v/\partial z) + Uv$$

generally, so giving:

First Telegrapher's Equation for $U \neq 0$; q_j may be finite or zero

$$\partial v/\partial z = -[(q/D) - (U/D)v] \qquad \text{for} \quad D \neq 0 \tag{6.1a}$$

or

$$0 = -[q - Uv] \qquad \text{for} \quad D = 0 \tag{6.1b}$$

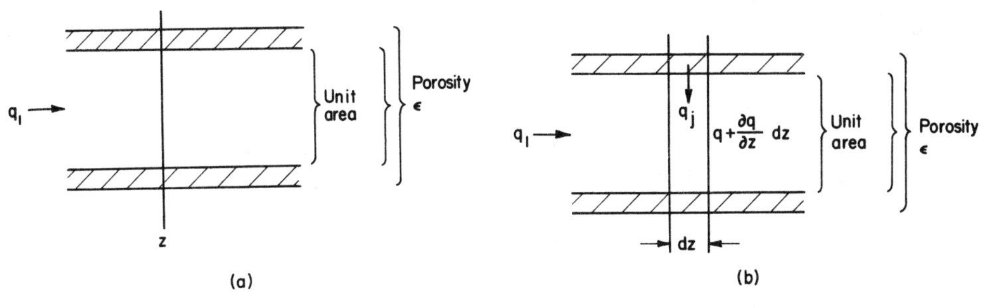

Fig. 6.3. Derivation of (a) first, and (b) second telegrapher's equations.

From Figure 6.3(b) we have for time dt and unit cross-sectional area

$$\underset{\text{In}}{q_1 \times 1\,dt + q_j \times 1\,dz\,dt} = \underset{\text{Out}}{[q_1 + (\partial q_1/\partial z)\,dz] \times 1\,dt} + \underset{\text{Accumulation}}{(\partial v/\partial t) \times 1\,dz\,dt}$$

so giving:

Second Telegrapher's Equation; U and q_j may be finite or zero

$$\partial q_1/\partial z = -[(\partial v/\partial t) - q_j] \tag{6.2}$$

The telegrapher's equations for many other situations, both electrical and (nonflow) physical, are set out in Moore [1].

If the telegrapher's equations (6.1) and (6.2) are combined, by differentiating the first with respect to z and then eliminating $\partial q/\partial z$ between them, the wave equations (1.5a)–(1.5d) are obtained. Appendix 8 contains further discussion.

6.3 IMPEDANCE

6.3.1 Introduction

Generally, impedance is the ratio of some kind of driving force to the consequent flux. It follows that it will have dimensions; it also follows that it must account for the time-varying attributes of both force and flux. It is used in traveling waves of all kinds, e.g., sound waves as well as in electrical systems. If the driving force varies sinusoidally with time, the flux will likewise vary with its own amplitude, the same frequency, and with a phase angle relative to the driving force. It is thus the ratio of two vectors and is itself a vector. Further, if the force and flux are functions of distance, then impedance, too, may be a function of distance. Thus, impedance could be written as

$$Z(z) = V/I \tag{6.3}$$

where V denotes the phasor of the driving force, and I denotes the phasor of the flux; it is, in fact, so written for electrical and other physical systems, but in the case of a flowing medium it is necessary to write impedances as

$$Z^+(z) = v^+/q^+ \tag{6.4}$$

and

$$Z^-(z) = v^-/q^- \tag{6.5}$$

for reasons that will become clear in the next section. It will also follow that for the linear systems discussed here the impedance is not a function of distance z.

6.3.2 Incident and Reflected Waves and Their Impedances

The general solution to the second-order wave equation (i.e., when $D \neq 0$) is

$$V = Ae^{\gamma_1 z} + Be^{\gamma_2 z} \qquad \text{(1.10a) R}$$

and when $D = 0$

$$V = Ae^{\gamma_1 z} \qquad \text{(1.10b) R}$$

The first may be thought of as a general case, the constant B in the special case, viz., Equation (1.10b), being zero.

A most fruitful approach, long used in other branches of traveling wave engineering and mentioned in Chapter 1, is to consider that V above is the phasor that represents a wave that is made up of the combination of two others—one traveling in the direction of z increasing (the incident or positive wave), the other in the opposite direction (the reflected, echo, or negative wave). The amplitudes of both waves decrease in the direction of travel, as will be shown, but whereas for other systems the rate of decrease is the same, in the case of a flowing medium these rates are different, i.e. $|\gamma_1| \neq |\gamma_2|$ in Equation (1.10a).

In general, for $U \neq 0$, $\gamma_1 = \alpha_1 - i\beta$ and $\gamma_2 = \alpha_2 + i\beta$, and α_1 and α_2 are the attenuation coefficients, and measure the rate at which the absolute value of the amplitude will change. α_1 always turns out to be negative, and α_2 to be positive, while β is always positive: see note below.

Now, if the instantaneous value of the concentration is written down from Equation (1.6), then

$$v = Ve^{i\omega t} = Ae^{\alpha_1 z}e^{i(\omega t - \beta z)} + Be^{\alpha_2 z}e^{i(\omega t + \beta z)} \tag{6.6}$$

From the real part there follows

$$v = Ae^{\alpha_1 z}\cos(\omega t - \beta z) + Be^{\alpha_2 z}\cos(\omega t + \beta z) \tag{6.7}$$

Now, from Equations (1.1) and (1.4) it can be seen that the two parts in Equation (6.7) would represent waves traveling in opposite directions, and from Chapter 1 it follows that both these waves are pure sinusoids, the frequency ω, wavelength $\lambda = 2\pi/\beta$, and the velocity $\omega\lambda/2\pi$ of both being the same, but the rate of attenuation per unit distance being different.

Thus, the solutions of the wave equations (1.5a)–(1.5d) will, in general, give rise to expressions which may be considered to be the positive and negative waves of concentration and flux. That is

$$V = Ae^{\gamma_1 z} + Be^{\gamma_2 z} \quad \text{Vector addition} \tag{6.8}$$

$$V = v^+ + v^- \quad \text{Vector addition} \tag{6.9}$$

Hence,

$$v^+ = Ae^{\gamma_1 z} = Ae^{\alpha_1 z}e^{-i\beta z} \tag{6.10}$$

$$v^- = Be^{\gamma_2 z} = Be^{\alpha_2 z}e^{i\beta z} \tag{6.11}$$

Then, by means of the first telegrapher's equation an expression for the flux may be obtained from these last equations. Finally, the ratio of concentration to flux will give the impedances; in a flowing medium these will be different, depending on whether the positive or the negative wave is considered. These will now be derived.

(i) Positive Waves and Impedance

$$v^+(z) = Ae^{\gamma_1 z} = Ae^{\alpha_1 z}e^{-i\beta z} \tag{6.10) R}$$

and by the first telegrapher's equation (6.1)

$$q^+(z) = Uv^+ - (D\,\partial v^+/\partial z) \tag{6.12}$$

Hence,

$$q^+(z) = A[U - D(\alpha_1 - i\beta)]e^{\alpha_1 z}e^{-i\beta z} \tag{6.13}$$

So,

$$Z^+ = v^+/q^+ = 1/[U - D(\alpha_1 - i\beta)] \tag{6.14}$$

(ii) Negative Waves and Impedance

$$v^-(z) = Be^{\gamma_2 z} = Be^{\alpha_2 z}e^{i\beta z} \tag{6.11) R}$$

and by the first telegrapher's equation (6.1)

$$q^-(z) = Uv^- - (D\,\partial v^-/\partial z) \tag{6.15}$$

Hence,

$$q^-(z) = B[U - D(\alpha_2 + i\beta)]e^{\alpha_2 z}e^{i\beta z} \tag{6.16}$$

So

$$Z^- = v^-/q^- = 1/[U - D(\alpha_2 + i\beta)] \qquad (6.17)$$

In order to find values of Z^+ and Z^- from Equations (6.14) and (6.17), it is necessary to know values of α_1, α_2, and β for particular cases, as will now be demonstrated. In so doing, the solutions to equations (1.5a)–(1.5d) will be needed for the steady cyclic state. However, when $q_j \neq 0$ it is necessary to replace this by a function of v_1. As Chapter 7 will demonstrate, the relation is conveniently expressed in terms of a *complex shunt admittance* $Y = Y_1 + iY_2$ of a reservoir phase that supplies the external flux such that

$$Q_j = -\tilde{V}_j Y_j V_1 \qquad \text{or} \qquad q_j = -\tilde{V}_j Y_j v_1 \qquad (7.3)\ \text{R}$$

and $\tilde{V}_j$ is the volume of the jth reservoir phase per unit volume of flowing phase (subscript 1).

If, further, there are n sources of external flux, all "in parallel," i.e., they act independently of each other, then everywhere q_j (etc.) may be replaced by $\sum_{j=2}^{n+1} q_j$ (etc.), but this will not ordinarily be written into the general equations.

By using Equation (7.3) the solutions of Equations (1.5a)–(1.5d) may now be derived.

Note

That α_1 is always negative may be seen from Equation (6.50) (for α_2) with the negative root taken. Everything under the root signs is positive, including Y_1 and Y_2 (see Chapter 7) and so the root is always numerically larger than $U/2D$.

Case a. $U \neq 0$, $D \neq 0$, $q_j \neq 0$ *[Equation (1.5a)]*

Combination of Equations (1.5a) and (7.3), together with the fact that $v = Ve^{i\omega t}$, gives

$$D(d^2V/dz^2) - U(dV/dz) - [i\omega + \tilde{V}(Y_1 + iY_2)]V = 0 \qquad (6.18)$$

(subscripts 1 and j being dropped).

If the idea of incident and reflected waves is used, then its solution gives

$$v^+ = Ae^{\gamma_1 z} \qquad (6.19)$$

$$v^- = Be^{\gamma_2 z} \qquad (6.20)$$

where

$$\gamma_1 = \alpha_1 - i\beta = (U/2D) - \{(U/2D)^2 + (1/D)[Y_1\tilde{V} + i(Y_2\tilde{V} + \omega)]\}^{1/2} \qquad (6.21)$$

$$\gamma_2 = \alpha_2 + i\beta = (U/2D) + \{(U/2D)^2 + (1/D)[Y_1\tilde{V} + i(Y_2\tilde{V} + \omega)]\}^{1/2} \qquad (6.22)$$

If new symbols are introduced for convenience, viz.,

$$b = (U/2D)^2 + (1/D)Y_1\tilde{V} \tag{6.23}$$

$$g = (1/D)(Y_2\tilde{V} + \omega) \tag{6.24}$$

$$s = +(b^2 + g^2)^{1/2} \tag{6.25}$$

then by De Moivre's theorem (Appendix 9) it follows that

$$\alpha_1 = (U/2D) - [(s + b)/2]^{1/2} \tag{6.26}$$

$$\alpha_2 = (U/2D) + [(s + b)/2]^{1/2} \tag{6.27}$$

$$\beta = [(s - b)/2]^{1/2} \tag{6.28}$$

and

$$\alpha_2 = \alpha_1 + 2[(s + b)/2]^{1/2} \tag{6.29}$$

In order to compute values of the above, it is necessary to have the sources of q_j (the *reservoir phases*) defined so that the shunt admittance $Y_1 + iY_2$ can be worked out. Some examples are given in Chapter 7.

Case b. ***$U \neq 0$, $D \neq 0$, $q_j = 0$ [Equations (1.5b) and (1.5b'')]***

By putting $Y_1 = Y_2 = 0$ into Equations (6.21) and (6.22), and by using De Moivre's theorem (Appendix 9), it follows that

$$\mathfrak{v}^+ = Ae^{\gamma_1 z} = Ae^{\alpha_1 z}e^{-i\beta z} \tag{6.30}$$

$$\mathfrak{v}^- = Be^{\gamma_2 z} = Be^{\alpha_2 z}e^{i\beta z} \tag{6.31}$$

where

$$\alpha_1 = (U/2D) - (U/2D)[(T + 1)/2]^{1/2} \tag{6.32}$$

$$\alpha_2 = (U/2D) + (U/2D)[(T + 1)/2]^{1/2} \tag{6.33}$$

$$\beta = (U/2D)[(T - 1)/2]^{1/2} \tag{6.34}$$

and

$$\alpha_2 = \alpha_1 + 2(U/2D)[(T - 1)/2]^{1/2} \tag{6.35}$$

where

$$T = (1 + W^2)^{1/2} \tag{6.36}$$

and

$$W = 4\omega D/U^2 \tag{6.37}$$

(Positive roots only are taken.)

Case c. ***$U \neq 0$, $D = 0$, $q_j \neq 0$ [Equation (1.5c)]***

This case (and case d) gives rise to a first-order differential equation [Equation (1.5c)], which, after substitution of Equation (7.3), gives

$$U(dV/dz) + [Y_1\tilde{V} + i(Y_2\tilde{V} + \omega)]V = 0 \tag{6.38}$$

from which it immediately follows that

$$v^+ = V = A[\exp(-Y_1\tilde{V}z/U)]\exp(-i(Y_2\tilde{V} + \omega)z/U) \quad (6.39)$$

as in Rosen and Winsche [2], and, further, that

$$v^- = 0 \quad (6.40)$$

i.e.,

$$\alpha_1 = -Y_1\tilde{V}/U \quad (6.41)$$

and

$$\beta = (Y_2\tilde{V} + \omega)/U \quad (6.42)$$

while α_2 is discussed below.

Case d. $U \neq 0$, $D = 0$, $q_j = 0$ [*Equation (1.5d)*]

For the steady cyclic state Equation (1.5d) becomes

$$U(dV/dz) + i\omega V = 0 \quad (6.43)$$

Hence

$$v^+ = V = Ae^{-i\omega z/U} \quad (6.44)$$

and

$$v^- = 0; \quad (6.45)$$

i.e.,

$$\alpha_1 = 0 \quad (6.46)$$

and

$$\beta = \omega/U \quad (6.47)$$

while α_2 is discussed below.

Impedance for the Above Four Cases

The expressions for α_1, α_2, and β given above [viz., case a: Equations (6.26), (6.27), (6.28); case b: Equations (6.32), (6.33), (6.34)] can now be substituted into Equations (6.14) and (6.17) to give expressions for Z^+ and Z^- (needed later) in these two cases, viz., when $D \neq 0$.

However, when D *is* 0, there is some uncertainty about the impedance to the reflected wave, Z^-, although the impedance for the incident wave comes out by putting $D = 0$ into Equation (6.14), viz.,

$$\text{Cases c, d:} \qquad Z^+ = 1/U \quad (6.48)$$

irrespective of what Y_1 and Y_2 are. But for Z^- the value of α_2 is required. Now, there is no reflected wave, i.e., $v^- = 0$. Hence, B is put equal to zero, but the value of α_2 is not immediately obvious (although β should, of course, be the same as for the incident wave). However, it can be shown (see the problem following) that:

$$\text{Cases c, d:} \qquad Z^- = \infty + i\infty \tag{6.49}$$

This fact is required when calculating reflection coefficients (below).

It will be seen that when $D = 0$ and $q_j = 0$ the wave velocity, ω/β, is independent of frequency, while when either $D \neq 0$ or $q_j \neq 0$ it is so dependent. In the language of traveling wave engineering the wave is *dispersive* when its velocity is dependent upon frequency; in the language of chemical engineering the system is *dispersive* when $D \neq 0$; in the language of chromatography there is *dispersion* when $q_j \neq 0$. Thus, by happy chance three meanings of one word in the jargons of three branches of knowledge have come together in a satisfying manner.

It will also be seen that when $D = 0$ the impedance in a positive direction is always $1/U$ and the impedance in a negative direction is always infinite. Furthermore, there is no negative wave, yet, as will be seen later there is reflection at a boundary. Thus, the reflected wave must die out in zero (upstream) distance.

Problem

Find α_2 and β (two ways) for the reflected wave when $D = 0$. Hence find Z^-.

Answer

(i) EXPRESSIONS FOR α_2 AND β

From the expression for α_2 in the general case, viz., Equation (6.27), it can be written as

$$\begin{aligned}\alpha_2 = (U/2D) + \{[(U^2/4D^2)^2 + (Y_1\tilde{V}/D)^2 + (2Y_1\tilde{V}U^2/4D^3)\\ + (Y_2\tilde{V} + \omega)^2/D^2]^{1/2} + (U/2D)^2 + (Y_1\tilde{V}/D)\}^{1/2}(1/\sqrt{2})\end{aligned} \tag{6.50}$$

Hence, as $D \longrightarrow 0$

$$\lim[\alpha_2]_{D\to 0} \longrightarrow \infty \tag{6.51}$$

Again, the equation (6.28) for β can be written as

$$\begin{aligned}\beta = \{[(U^2/4D^2)^2 + (Y_1\tilde{V}/D)^2 + (2Y_1\tilde{V}U^2/4D^3) + (Y_2\tilde{V} + \omega)^2/D^2]^{1/2}\\ - (U/2D)^2 + (Y_1\tilde{V}/D)\}^{1/2}(1/\sqrt{2})\end{aligned} \tag{6.52}$$

By a little algebraic manipulation it is found that

$$\lim[\beta]_{D\to 0} = (Y_2\tilde{V} + \omega)/U \tag{6.53}$$

as it should [Equation (6.42)].

(ii) THE VALUE OF Z^-

By Equation (6.17),

$$Z^- = 1/[U - D(\alpha_2 + i\beta)]$$

Now, as $D \longrightarrow 0$ the value of $D\beta$ tends to 0, but to obtain the limiting value of $D\alpha_2$ [other than the indeterminate 0/0 that the value of Part (i) gives], the expression for α_2 written as

$$\alpha_2 = \frac{U}{2D} + \frac{1}{D\sqrt{2}}$$
$$\times \left[\left(\frac{U^4}{16} + Y_1{}^2\tilde{V}^2D^2 + \frac{2U_2Y_1\tilde{V}D}{4} + Y_2{}^2\tilde{V}^2D^2 + \omega^2D^2 + 2Y_2\tilde{V}\omega D^2\right)^{1/2}\right.$$
$$\left. + \left(\frac{U^2}{4} + Y_1\tilde{V}D\right)\right]^{1/2} \tag{6.54}$$

Hence,

$$\lim[Z^-]_{D\to 0} = \lim\left[\frac{1}{U - D\alpha_2 - iD\beta}\right]_{D\to 0} = \frac{1}{U - U - i0}$$

i.e.,

$$\lim[Z^-]_{D\to 0} = \infty + i\infty \tag{6.55}$$

Note

The dimensions of impedance depend upon the system, as will be seen from its definition, e.g., Equations (6.4) and (6.5). If v is a concentration (heat or mass), then they will be [TL^{-1}]. If v is a temperature, then they will be [TL^2] [temperature/ heat]. However, the impedances will be used here (to find a reflection coefficient) in the form of a ratio.

6.4 FLUXES

Equation (6.13) gives the value of q^+ in the general case; similarly Equation (6.16) gives q^-, and values can be found if the terms in the equation are known. Note the following special cases.

Case c. $D = 0,\ q_j \neq 0$

$$q^+ = AU\exp\{-(1/U)[Y_1\tilde{V} + i(Y_2\tilde{V} + \omega)]z\} \tag{6.56}$$
$$q^- = 0 \tag{6.57}$$

Case d. $D = 0,\ q_j = 0$

$$q^+ = AUe^{-i\omega z/U} \tag{6.58}$$
$$q^- = 0 \tag{6.59}$$

6.5 REFLECTIONS AND REFLECTION COEFFICIENTS [3]

6.5.1 Introduction

In a noninfinite system there will, by definition, be a boundary; the medium must flow through this boundary, but in the general case the wave must satisfy Equation (1.10a) with $B \neq 0$. Thus, at the upstream side of any boundary the situation is considered to be that a positive wave $\mathfrak{v}^+$ reaches the boundary and a negative wave $\mathfrak{v}^-$ is reflected from it normally; both are being attenuated in the direction of travel.

The ratio of the reflected wave to the incident wave is termed the reflection coefficient Γ,

$$\Gamma = \mathfrak{v}^- / \mathfrak{v}^+ \tag{6.60}$$

and it is complex.

The above has been based on the assumption that the flux has been eliminated between the first and second telegrapher's equations, but it might as easily have been v, and so the waves and reflection coefficient would then have been expressed in terms of flux.

6.5.2 Behavior at a Boundary

(*i*) *Conditions*

Consider boundary *ab* between regions *a* and *b* in Figure 6.1. These subscripts designate the appropriate quantities.

Subject to the conditions and discussion given at the beginning of Chapter 1, it may be specified that at the boundary, since there is no discontinuity of temperature or fugacity and no accumulation (i.e., flux in = flux out), then the boundary conditions, analogous to those at a terminal or junction in electrical theory, are

$$\left.\begin{aligned} v_a &= v_b \qquad (6.61)\\ A_a q_a &= A_b q_b \qquad (6.62) \end{aligned}\right\} \text{ at the boundary}$$

Since

$$v_a = V_a e^{i\omega t} = v_b = V_b e^{i\omega t}$$

and

$$A_a q_a = A_a Q_a e^{i\omega t} = A_b q_b = A_b Q_b e^{i\omega t}$$

it follows that at the boundary

$$V_a = V_b \tag{6.63}$$

$$A_a Q_a = A_b Q_b \tag{6.64}$$

Furthermore, if Equation (6.62) and the first telegrapher's equation, viz. (6.1a), are compared, it will be seen that

$$U_a A_a v_a - D_a A_a \, \partial v_a/\partial z = U_b A_b v_b - D_b A_b \, \partial v_b/\partial z \tag{6.65}$$

which is the same as the one discussed in Chapter 5; i.e., Equations (6.61) and (6.62) are Wehner and Wilhelm conditions, but for time-varying concentrations.

(ii) Reflection and Transmission

[It is convenient to measure distance *upstream* from the boundary, designating this by d (positive) as in Figure 6.4, page 108.]

If the reflection coefficient at the boundary is known, then, by Equation (6.60), the relation between $\mathfrak{v}^+$ and $\mathfrak{v}^-$ at the boundary can be found; so far, however, these last quantities are only known in terms of the arbitrary constants A and B [in Equations (6.10) and (6.11)]. Now, the reflection coefficient can be found by using the above boundary conditions and the definition of impedance, in the following way.

At the Boundary

At the boundary $d = 0$, and since this is a significant location, Table 6.1 lists the following:

$$\mathfrak{v}^+(d = 0) \equiv V^+ \tag{6.66}$$

$$\mathfrak{v}^-(d = 0) \equiv V^- \tag{6.67}$$

and the vector sum,

$$V_R = V^+ + V^- \tag{6.68}$$

Similarly,

$$Q_R = Q^+ + Q^- \tag{6.69}$$

i.e.,

$$(V^+/Z^+) + (V^-/Z^-) = Q_R \tag{6.70}$$

(If Z^+ and Z^- were functions of z, then the values at $d = 0$ would have to be indicated in some way.)

Now, an "effective" or "receiving" impedance at $d = 0$ is

$$Z_R = \frac{V_R}{Q_R} = \frac{V^+ + V^-}{(V^+/Z^+) + (V^-/Z^-)} \tag{6.71}$$

But, by Equations (6.60), (6.66), and (6.67) $\Gamma = V^-/V^+$. Hence, Equation (6.71) gives $Z_R = (1 + \Gamma)/[(1/Z^+) + (\Gamma/Z^-)]$, or, rearranged

$$\Gamma = \frac{Z^-}{Z^+}\left[\frac{Z_R - Z^+}{Z^- - Z_R}\right] \tag{6.72}$$

Now, Z^+ and Z^- may be found from the earlier expressions (Section 6.3.2) in this chapter, while Z_R is simply related to conditions downstream of the boundary, as can be shown from Equations (6.63) and (6.64). For, $V_a =$ resultant V_R at $d = 0$ by Equation (6.68), and $Q_a =$ resultant Q_R at $d = 0$ by Equation (6.69), while

$$Z_R = \frac{V_R}{Q_R} = \frac{V_a}{Q_a} = \frac{V_b}{A_b Q_b / A_a}$$

But, if

$$Z_b = V_b / Q_b \tag{6.73}$$

then

$$Z_R = A_a Z_b / A_b \tag{6.74}$$

Substitution of the last equation into Equation (6.72) gives

$$\Gamma = \frac{Z_a^-}{Z_a^+}\left[\frac{(A_a/A_b)Z_b - Z_a^+}{Z_a^- - (A_a/A_b)Z_b}\right] \tag{6.75}$$

Also, by Equation (6.68), $V_R = V_a = V^+ + V^-$;

$$\therefore \quad V_R/V^+ = (V^+ + V^-)/V^+ = 1 + (V^-/V^+) = 1 + \Gamma \tag{6.76}$$

Hence, if the distributions of amplitude and phase angle with distance is required at any point in a region upstream of a boundary, they may be obtained by combining $v^+(z)$ and $v^-(z)$ vectorially. It should be remembered that $v^+(z)$ is the same as would obtain if the boundary were removed to infinity, while v^- may be obtained if the reflection coefficient is known. Now, the reflection coefficient can be calculated by Equation (6.75) if the appropriate impedances are known. These in turn may be obtained by means of Equations (6.14) and (6.17).

Finally, the details of the transmitted wave at the beginning of the downstream section can be obtained from Equation (6.76) as

$$V_R = V^+(1 + \Gamma)$$

6.5.3 Resulting Situation near a Boundary

As stated above, it is convenient to use d, the distance upstream from a boundary at $z = L$. In terms of this, the equations for the incident and reflected wave become, from Equations (6.10) and (6.11) and the fact that $z = L - d$,

$$v^+ = Ae^{\gamma_1 z} = Ae^{\gamma_1(L-d)} = Ae^{\gamma_1 L}e^{-\gamma_1 d}$$

i.e.,

$$v^+ = V^+ e^{-\gamma_1 d} = V^+ e^{-\alpha_1 d} e^{i\beta d} \tag{6.77}$$

Similarly,

$$\mathcal{v}^- = V^- e^{-\gamma_2 d} = V^- e^{-\alpha_2 d} e^{-i\beta d} \tag{6.78}$$

(*N.B.:* There is a slight difference between the above nomenclature and that in transmission line theory, where γ signifies the absolute value, and positive and negative signs are attached. In the present case, however, γ_1 and γ_2 may be intrinsically positive or negative.)

Thus, near a boundary, on the upstream side, we have the situation shown in Figure 6.4.

The incident wave described by Equation (6.77) is traveling from left to right, the phasors at intervals of a quarter of a wavelength being shown as full lines. The reflected wave described by Equation (6.78) after its magnitude at the boundary has been found by using Γ is thus traveling from right to left, its phasors being shown by dashed lines. The phasors of the two waves rotate in opposite-handed ways. For example, if the incident phasor rotates as a right-handed screw thread, then the reflected phasors revolve as a left-handed thread. At any given distance from the boundary the resultant phasor, giving the amplitude and phase angle of the effective wave at that point, may be found by combining the phasors vectorially as Figure 6.4. (Figure 6.4 also indicates that distance d is measured upstream from the boundary.)

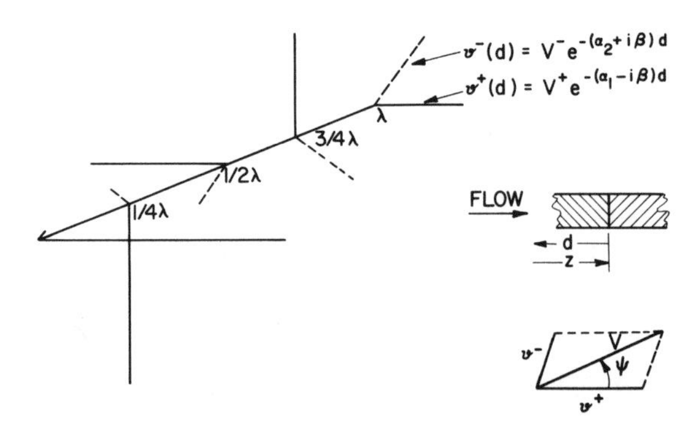

Fig. 6.4. Phasors of an incident (full) and reflected (dashed) sine wave at various distances.

The above expressions for the reflection coefficient have assumed that no reflected wave from a *subsequent* boundary has reached the one in question. This is a fairly safe assumption; the reflected wave is attenuated so rapidly in the situation under review that were it not so, the system would have to be so short that the basic assumptions would be in doubt. However, if it did, the reflection coefficient can still be found for the new case (see Moore [1]).

6.6 COMPENSATION FOR END EFFECTS

In a determination of parameters by measuring frequency response, of necessity an infinite system cannot be used, and so in general two boundaries exist in the apparatus, and these, along with the measuring and detecting devices themselves, affect the measured values. It is not necessary to calculate the effects of boundaries and measuring devices—fortunately, for even if this were done, the simplicity of the change with distance of the amplitude (exponential) and of the phase angle (linear) that occurs in an infinite bed would be lost. The procedure is to find the responses at two bed lengths [4–7] l_1 and l_2 (Figure 6.5). Then, the differences of the logarithm of the

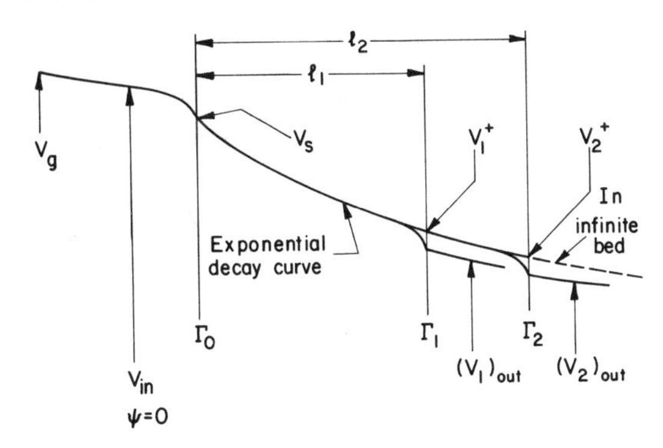

Fig. 6.5. Compensation for end effects. Diagrammatic representation of the resultant phasor.

amplitude ratio and of the phase angle in the two cases give the true response of a bed of length $l_2 - l_1$, provided that the effects of the boundary are as shown in Figure 6.5, i.e., they do not extend far enough upstream to affect either the first boundary or the line indicating length l_1. The justification can now be given.

The situation is shown in Figure 6.5; the line that indicates the amplitude is diagrammatic (for the amplitude is a vector), but it serves to locate the magnitudes in the argument. Basically, a sinusoidal perturbation of temperature or concentration V_g is generated upstream, a value (modified by the response of the detector system—and this could be formally incorporated into the following argument) V_{in} is measured before the bed and again downstream as V_{out}. At the entrance and exit of the bed there is reflection of part of the incident wave, the reflected wave being rapidly attenuated. The actual but unknown inlet (or sending) amplitude is V_s; the amplitude that would exist at the exit of the bed if the simple "infinite bed" law of attenuation held would be V^+. For two beds of length l_1 and l_2 it is required to find V_1^+/V_s and V_2^+/V_s, thus giving V_2/V_1 for a bed of length $l_2 - l_1$; however, the quantities actually measured are $(V_1)_{out}/V_{in}$ and $(V_2)_{out}/V_{in}$, but with the provisos listed below, the two last quantities will give the desired ones. For, generally, $V_{out} \propto V^+(1 + \boldsymbol{\Gamma})$, so

$$(V_m)_{out}/V_{in} = V_m^+(1 + \boldsymbol{\Gamma}_m)V_s/V_sV_{in} = (V_m^+/V_s)(1 + \boldsymbol{\Gamma}_m)(1 + \boldsymbol{\Gamma}_0), \quad m = 1, 2$$

from which it follows that if $\boldsymbol{\Gamma}_1 = \boldsymbol{\Gamma}_2$,

$$[(V_2)_{out}/V_{in}]/[(V_1)_{out}/V_{in}] = [V_2^+/V_s]/[V_1^+/V_s]$$

i.e.,

$$\frac{\text{Measured}}{[|V_2|_{out}/V_{in}]/[|V_1|_{out}/V_{in}]} = \frac{\text{Desired}}{[|V_2^+|/|V_s|]/[|V_1^+|/|V_s|]}$$

$$(\psi_2)_{out} - (\psi_1)_{out} = \psi_2^+ - \psi_1^+$$

The last relations are, of course, those that have been used; it will now be seen that the conditions under which they are true are:

(1) $l_2 - l_1$ is sufficiently large for the reflected wave from boundary 2 not to reach boundary 1 to any appreciable extent.

(2) l_1 is sufficiently large for the reflected wave from boundary 1 not to reach boundary 0.

(3) $\boldsymbol{\Gamma}_1 = \boldsymbol{\Gamma}_2$.

(4) The change in complex amplitude between the entrance or exit boundary and the corresponding measuring device is zero or constant for all values of l.

(5) The responses of the measuring devices are unaltered by varying l.

Corrections

Conditions (2) and (3) in the last paragraph render doubtful the method of correcting for end effects by measuring the response of the system with the active part removed; this was practiced by McHenry and Wilhelm [6] (determination of a longitudinal dispersion coefficient), Turner [7] (analysis of the flow structures of packed beds), and Goss [4] (determination of thermal parameters), although the latter could find no significant difference between this method and the one involving two lengths of the active bed; the corrections were small. The magnitude of the errors may be found by the method (involving reflection coefficients) described by Turner [3].

6.7 SPECIAL BUT FREQUENTLY OCCURRING CASE

As was shown earlier, it happens that when $D = 0$ then $Z^+ = 1/U$ and $Z^- = \infty + i\infty$. In that case we have the following expressions for the reflection coefficients for the two situations in Figure 6.6.

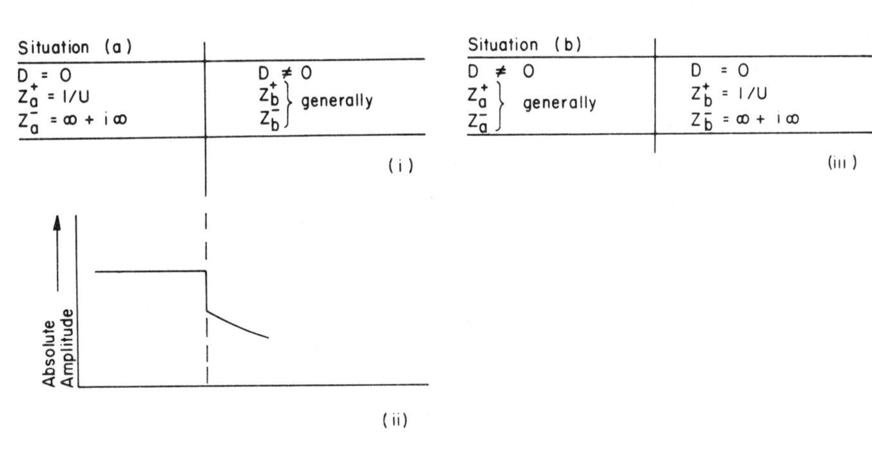

Fig. 6.6. Reflection coefficients at a boundary and impedances in each region.

Reflection Coefficient for Situation (a). Figure 6.6(i)

$$\Gamma_a = (A_a/A_b)UZ_b^+ - 1 \tag{6.79}$$

If, further, $A_a = A_b = 1$ and Equation (6.14) is applied then

$$\Gamma_a = \frac{(D\alpha_1/U)[1 - (D\alpha_1/U)] - (D\beta/U)^2 - i(D\beta/U)}{(1-D\alpha_1/U)^2+(D\beta/U)^2} \tag{6.80}$$

Reflection Coefficient for Situation (b). Figure 6.6(iii)

$$\Gamma_b = \frac{Z_a^-}{Z_a^+}\left[\frac{(A_a/A_b)(1/U) - Z_a^+}{Z_a^- - (A_a/A_b)(1/U)}\right] \tag{6.81}$$

If, further, $A_a/A_b = 1$ and Equations (6.14) and (6.17) are applied, then

$$\Gamma_b = -\left[\frac{\alpha_1 - i\beta}{\alpha_2 + i\beta}\right]$$

$$= -\left[\frac{\alpha_1\alpha_2 - \beta^2 - i\beta(\alpha_2 + \alpha_1)}{\alpha_2^2 + \beta^2}\right] \tag{6.82}$$

Discussion

It will be seen from Equation (6.79) that in the situation of Figure 6.6(i) there is a finite reflection coefficient (albeit small, as the following examples will indicate), but it was implied by Section 6.3 that there is no reflected wave in the upstream section, and no attenuation either. That means that the *amplitudes* of the wave upstream and downstream of the boundary would be as in Figure 6.6(ii); that is, there is the paradox of an abrupt change at this boundary and yet no discontinuity of concentration, as demanded by Equation (6.63). This is analogous to the steady-state concentration profile at the entrance to a reactor, as pointed out by Wehner and Wilhelm [8]; on the other hand, at the exit boundary, i.e., in the case illustrated in Figure 6.6(iii), there is no real analogy. This question was discussed in Chapter 5.

Note

The right-hand sides of Equations (6.80) and (6.82) could be further developed by using Equations (6.26)–(6.28), but it is just as easy to put in numerical values for α_1, α_2, and β in any quantitative problems. Some are given below: in these, both cross-sectional areas and the velocities are the same in all regions, but no difficulty arises if they are not.

6.8 WORKED EXAMPLES

Problem 1. Figure 6.7(a)

Find the longitudinal impedances and hence the reflection coefficients for the system shown in Figure 6.7(a) given that $r_0 = 0.1$ cm, $\omega = 1.0$ rad sec^{-1}, $U = 2500$ cm sec^{-1}, $D = 250$ cm^2 sec^{-1}, obtained by assuming that the particle Peclet number $2Ur_0/D = 2$.

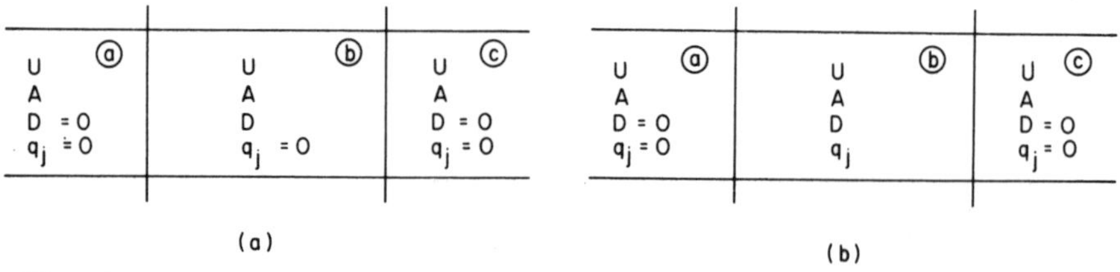

Fig. 6.7. Worked examples. (Section 6.8.) (a) System of Problem 1. (b) System of Problem 2.

Answers

Shunt Admittances

$Y = 0$, since $q_j = 0$.

Propagation Constants in Section b

$U/2D = 2500/(2 \times 250) = 5.0.$

$$W = 4\omega D/U^2 = [4\omega(2r_0)/U][D/2Ur_0] \qquad \text{[Equation (6.37)]}$$
$$= (4 \times 1.0 \times 0.2/2500) \times \tfrac{1}{2} = 1.6 \times 10^{-4}$$
$$T = (1 + W^2)^{1/2} \qquad \text{[Equation (6.36)]}$$
$$= (1 + 2.56 \times 10^{-8})^{1/2} \approx 1 + 1.28 \times 10^{-8}$$
$$\alpha_1 = (U/2D)\{1 - [(T+1)/2]^{1/2}\} \qquad \text{[Equation (6.32)]}$$
$$= 5\{1 - [(2 + 1.28 \times 10^{-8})/2]^{1/2}\}$$
$$\approx 5[1 - (1 + 0.32 \times 10^{-8})]$$
$$= -1.6 \times 10^{-8}$$
$$\alpha_2 = (U/2D)\{1 + [(T+1)/2]^{1/2}\} \qquad \text{[Equation (6.33)]}$$
$$= 5(1 + 1 + 0.32 \times 10^{-8})$$
$$= 10 + 1.6 \times 10^{-8}$$
$$\beta = (U/2D)[(T-1)/2]^{1/2}\} \qquad \text{[Equation (6.34)]}$$
$$= 5[(1 + 1.28 \times 10^{-8} - 1)/2]^{1/2}$$
$$= 4.0 \times 10^{-4}$$

Impedances

SECTION a

$$Z_a^+ = 1/U \qquad \text{[Equation (6.48)]}$$
$$= 4 \times 10^{-4} \quad \text{sec cm}^{-1}$$
$$Z_a^- = \infty + i\infty \qquad \text{[Equation (6.55)]}$$

SECTION b

$$Z_b^+ = 1/\{U[1 - (D/U)(\alpha_1 - i\beta)]\} \qquad \text{[Equation (6.14)]}$$
$$= 1/\{2500[1 - 0.1(-1.6 \times 10^{-8} - 4.0 \times 10^{-4}i)]\}$$
$$= (1/2500)(1 - 16.0 \times 10^{-10} - 4.0 \times 10^{-5}i)$$
$$Z_b^- = 1/\{U[1 - (D/U)(\alpha_2 + i\beta)]\} \qquad \text{[Equation (6.17)]}$$
$$= -(1/2500)(1 - 2.5 \times 10^{-4}i)$$

SECTION c

$$Z_c^+ = 1/U \qquad \text{[Equation (6.48)]}$$
$$= 1/2500$$
$$Z_c^- = \infty + i\infty \qquad \text{[Equation (6.55)]}$$

Reflection Coefficients

These may be found most conveniently by using Equations (6.79) and (6.82).

BOUNDARY *ab*

$$\Gamma_{ab} = UZ_b^+ - 1 \qquad \text{[Equation (6.79)]}$$
$$= (2500/2500)(1 - 16.0 \times 10^{-10} - 4.0 \times 10^{-5}i) - 1$$

i.e.,

$$\Gamma_{ab} = -16.0 \times 10^{-10} - 4.0 \times 10^{-5}i$$

BOUNDARY *bc*

$$\Gamma_{bc} = -\left[\frac{\alpha_1 - i\beta}{\alpha_2 + i\beta}\right] \qquad \text{[Equation (6.82)]}$$
$$= -\left[\frac{-1.6 \times 10^{-8} - 4.0 \times 10^{-4}i}{10 + 1.6 \times 10^{-8} + 4.0 \times 10^{-4}i}\right]$$
$$= 32 \times 10^{-10} + 4.0 \times 10^{-5}i$$

Diagrams (not to scale) of Γ_{ab} and Γ_{bc} are given in Figure 6.8(a) for Problem 1.

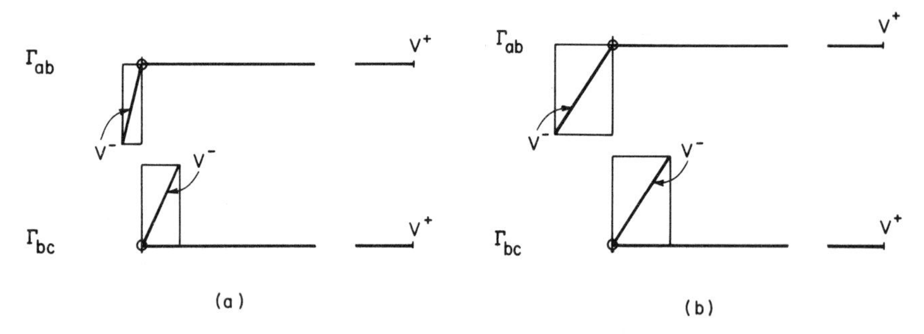

Fig. 6.8. Worked examples (Section 6.8). (a) Answer to Problem 1. (b) Answer to Problem 2.

Problem 2. Figure 6.7(b)

The problem is the same as Problem 1. Values of parameters are as in Problem 1, together with the following information about the source of q_j:

The source is a single reservoir phase; i.e., $j = \#2$.
Geometry: spherical.
Heat transfer coefficient $h = 25.0 \times 10^{-3}$ cal cm^{-2} sec^{-1} °C^{-1}.
Conductivity $k = 2.2 \times 10^{-3}$ cal sec^{-1} cm^{-1} °C^{-1}.
Volumetric specific heat $F_2 = 0.5$ cal cm^{-3} °C^{-1}.

Answers

(Computation of the propagation constants requires values of the shunt admittance of this particular reservoir phase, discussed at length in Chapter

7. These tedious calculations have been done by computer and described by Goss [4]. His values for this case are $\alpha_1 = -0.754$, $\beta = +0.802$.)

$$\begin{aligned}
\alpha_1 &= -0.754 \\
\alpha_2 &= 2(U/2D) - \alpha_1 \qquad \text{(always true)} \\
&= 2 \times 5 + 0.754 = 10.754 \\
\beta &= 0.802
\end{aligned}$$

Impedances

SECTIONS a AND c

The impedances in these sections are as in Problem 1.

SECTION b

$$\begin{aligned}
Z_b^+ &= 1/\{U[1 - (D/U)(\alpha_1 - i\beta)]\} \qquad \text{[Equation (6.14)]} \\
&= 1/\{2500[1 - 0.1(-0.754 - 0.802i)]\} \\
&= (1/2500)[0.926 - 0.0689i] \\
Z_b^- &= 1/\{U[1 - (D/U)(\alpha_2 + i\beta)]\} \qquad \text{[Equation (6.17)]} \\
&= 1/\{2500[1 - 0.1(10.75 + 0.802i)]\} \\
&= (1/2500)[-6.18 + 6.56i]
\end{aligned}$$

Reflection Coefficients

(See remark in Problem 1.)

BOUNDARY *ab*

$$\begin{aligned}
\Gamma_{ab} &= UZ_b^+ - 1 \qquad \text{[Equation (6.79)]} \\
&= (2500/2500)(0.926 - 0.069i) - 1
\end{aligned}$$

i.e.,

$$\Gamma_{ab} = -0.074 - 0.069i$$

BOUNDARY *bc*

$$\begin{aligned}
\Gamma_{bc} &= -\left[\frac{\alpha_1 - i\beta}{\alpha_2 + i\beta}\right] \qquad \text{[Equation (6.82)]} \\
&= -\left[\frac{-0.754 - 0.802i}{10.754 + 0.802i}\right]
\end{aligned}$$

i.e.,

$$\Gamma_{bc} = 0.0755 + 0.069i$$

Diagrams (not to scale) of Γ_{ab} and Γ_{bc} are given in Figure 6.8(b) for Problem 2.

Discussion of Problems

Reflected Waves

At boundary *ab* the reflected wave lagged the incident wave. When $q_j = 0$, the reflected wave was very small and the epoch was about $\pi/2$. When $q_j \neq 0$, the reflected wave was much larger (but still fairly small compared with the incident wave) and the epoch was about $3\pi/4$.

At boundary *bc* the phasors were about the same length, relative to the incident wave, as at *ab* but the reflected wave led the incident wave. When $q_j = 0$, the epoch was about $\pi/2$, when $q_j \neq 0$ the epoch was about $\pi/4$.

Wavelengths and Wave Velocities

In sections *a* and *c*, as well as in section *b* when there was no reservoir phase, the wave velocity ω/β was respectively equal to or very close to the fluid velocity 2500 cm sec^{-1}. The wavelength was therefore $\lambda = 2\pi/\beta = 1.57 \times 10^4$ cm.

On the other hand, when there was a reservoir phase in section *b* the wavelength and velocity there were markedly reduced, becoming $\lambda = 2\pi/0.802 = 7.85$ cm and $\omega/\beta = 1.0/0.802 = 1.25$ cm sec^{-1}, respectively.

6.9 THE INFINITE BED

As stated, even when of necessity measurements must be made on a bed of finite length, it is possible to get the harmonic response over a length of bed that behaves as though it were infinite, and with inlet and outlet concentrations precisely known. In this case there results $B = 0$ (otherwise, the wave would not be finite as $z \longrightarrow \infty$) and so

$$\mathscr{v}^+ = V^+ e^{\gamma_1 d}$$

When $d = -L$, then

$$\mathscr{v}^+(0) = V_s \qquad \text{(definition)}$$

Hence,

$$V^+ = V_s e^{\gamma_1 L}$$

L being measured in the positive direction.

The last equation gives rise, in all cases, to "nice" functions of distance for amplitude and phase angle, with the consequent promise of providing means of determining parameters in an elegant way. Even flow with velocity profiles that are not flat—e.g., Poiseuille flow—may be accommodated, with conditions, as mentioned in Appendix 10. Some attempts that have been made at parameter determination will be discussed briefly in Chapter 9.

6.10 THE "DANCKWERTS BOUNDARY CONDITION"

The boundary condition $\partial v/\partial z = 0$ is sometimes used at the exit from a bed as discussed in Section 5.1; it is valid only if $D = 0$ downstream and there is no change in cross sectional area, as will now be shown for sine waves. Other waves can be resolved into these.

At any value of z

$$v = \mathfrak{v}^+ + \mathfrak{v}^-$$

and as in Figure 6.4 or from Equations (6.77) and (6.78)

$$v = V^+e^{-\gamma_1 d} + V^-e^{-\gamma_2 d}$$

so

$$-\frac{\partial v}{\partial d} = \frac{\partial v}{\partial z} = 0 = +V^+\gamma_1 e^{-\gamma_1 d} + V^-\gamma_2 e^{-\gamma_2 d}$$

and at the exit, $d = 0$;
so

$$0 = +V^+\gamma_1 + V^-\gamma_2$$

or

$$\frac{V^-}{V^+} \equiv \Gamma = -\frac{\gamma_1}{\gamma_2} = -\left[\frac{\alpha_1 - i\beta}{\alpha_2 + i\beta}\right],$$

which is the value derived in Equation (6.82) for the case $A_a/A_b = 1$ and $D = 0$ downstream of the boundary shown in Figure 6.6(iii).

It is of interest to contrast the consequences of imposing the boundary condition $\partial v/\partial z = 0$ for the flow and nonflow cases. By the first telegrapher's equation, the longitudinal flux is given by $q = -D(\partial v/\partial z) + Uv$
If the condition $\partial v/\partial z = 0$ is applied, then

$$q = Uv \qquad \text{for the flow case, but}$$
$$q = 0 \qquad \text{for the nonflow case.}$$

In the latter case, where there is no flow beyond the boundary the receiving impedance $Z_R \longrightarrow \infty$, and the concentration reflection coefficient by Equation (6.72) is

$$\Gamma = -(Z^-/Z^+)$$

Now for the non flow case the system is reciprocal as, for example, in transmission line systems; in the latter case Z_0 is used for the numerical value of the intrinsic impedance and the negative sign is attached when dealing with negative-going waves, i.e., $Z_0 = Z^+ = Z^-$. In this book however Z^+ and Z^- keep their appropriate signs, and so, from Moore ([1], p. 67),

$$Z^+ = -Z^- = +(1/2a\omega)^{1/2}(1 - i)$$

and so $\Gamma = 1$ for concentration waves. (It is equal to -1 for flux waves.) Hence there is a standing wave with an antinode of concentration and a node of flux at the boundary; this is the situation at an insulated face of (nonflowing) material or at the center of a symmetrical system.

For both flow and non-flow systems the condition $\partial v/\partial z = 0$ requires that the gradients of both amplitude and phase-angle of concentration with distance be zero. For in general,

$$v = |V(z)|\, e^{i[\omega t - \psi(z)]}$$

and so for $\partial v/\partial z = 0$ for all time

$$0 = \left[\frac{\partial |V(z)|}{\partial z} - i|V(z)|\frac{\partial \psi(z)}{\partial z}\right] e^{-i\psi(z)} e^{i\omega t}$$

Thus both real and imaginary parts must be zero; that is, when $\partial v/\partial z = 0$, both $\partial |V|/\partial z = 0$ and $\partial \psi/\partial z = 0$ simultaneously.

REFERENCES

[1] R. K. Moore, "Traveling Wave Engineering." McGraw-Hill, New York, 1960.

[2] J. B. Rosen and W. E. Winsche, The Admittance Concept in the Kinetics of Chromatography. *J. Chem. Phys.* **18**, 1587 (1950).

[3] G. A. Turner, Reflection and Transmission of Thermal Waves at the Boundaries of a Fixed-Bed Regenerator. *Ind. Eng. Chem. Fundam.* **10**, 400 (1971).

[4] M. J. Goss, Determination of Thermal Parameters of a Packed Bed of Spheres by a Method of High Precision Frequency Response. Ph.D. Thesis, Univ. of Waterloo (1969).

[5] A. W. Liles and C. J. Geankoplis, Axial Diffusion of Liquids in Packed Beds and End Effects. *AIChE J.* **6**, 591 (1960).

[6] K. W. McHenry, Jr. and R. H. Wilhelm, Axial Mixing of Binary Gas Mixtures Flowing in a Random Bed of Spheres. *AIChE J.* **3**, 83 (1957).

[7] G. A. Turner, The Frequency Response of Some Illustrative Models of Porous Media. *Chem. Eng. Sci.* **10**, 14 (1959).

[8] J. F. Wehner and R. H. Wilhelm, Boundary Conditions of Flow Reactor. *Chem. Eng. Sci.* **6**, 89 (1956).

Chapter 7

Reservoir Phases

7.1 INTRODUCTION

There was a term q_j meaning a flux from a phase other than phase 1 (the flowing one), in some of the differential equations in the earlier chapters. If this flux comes from a material phase (rather than from radiation or chemical reaction) in communication with the flow, then this phase, called a reservoir, will have its own behavior that has to be considered when temperature or concentration is time-varying. This means that the phase has to be described, and its behavior investigated, in mathematical terms, and in order to proceed, the time variable must be replaced, as it has been in the foregoing, by s, the Laplace variable, or $i\omega$. Since these may be interchanged (as demonstrated in texts on the Laplace transform), the same procedure may be used in both cases; that is to say, a partial differential equation in time and spatial dimension(s) is set up that describes the reservoir phase. Transformation of t by s or $i\omega$ results in an equation that may be substituted into the main equation of the flowing medium. The similarity between the treatment using s or $i\omega$ tends now to end, for the use, interpretation, and (particularly) the experimental requirements now diverge. Hence, the logic of the two procedures will be discussed separately.

7.2 THE STEADY CYCLIC STATE

As the medium flows past the reservoir, its concentration showing a steady cyclic variation, it will cause heat or material to flow in or out of the reservoir. If the system is assumed to be linear, then this flux will have the same frequency as the concentration in the flowing medium, its magnitude will be related to the value of the concentration, and there will, in general, be a phase-angle difference between the two waves. Thus, the relation between the two will be a vector one. If q_j and v_1 are instantaneous values of flux and concentration, respectively, (see Chapter 6), and Q_j and V_j the phase vectors, then

$$v_1 = V_1 e^{i\omega t} \quad \text{and} \quad q_j = Q_j e^{i\omega t}$$

To discuss the relation between Q_j and V_1 further, the definition of Rosen and Winsche [1] is used; as stated in Chapter 6, the *impedance* of a system is a measure of the difficulty with which a flow occurs when a driving force is applied. The best known example is in electrical circuits, where the flow is of electricity and the driving force is the emf.

If Figure 1.4 is studied, it will be seen that the quantities L and R are connected in such a way that they will influence the flow of current along the system, but the quantities G and C will permit current to be *shunted* or bypassed, from one conductor to the other. If the amount so shunted is designated ΔI, the relation between this and the emf between the conductors at any point can be written as

$$\mathbf{\Delta I} = \mathbf{YE}\, dl$$

Bold type is used here to indicate that there would be a vector relation between them with a cyclic emf applied, but in general the use of bold type is dropped unless it is needed to avoid confusion.

The quantity $\mathbf{Y}$ is called a *shunt admittance per unit length.* The larger it is, the greater is the shunted current. [In the transmission line of Figure 1.4 its value is $\mathbf{Y} = G + i\omega C$.]

In our present physicochemical case the heat or matter moving "sideways" into the reservoirs may be considered to be analogous to this shunted electrical current, and the relation between this flux and the concentration may be termed the shunt admittance **Y**. If this is defined by: Rate of change of mean *concentration* in the reservoir phase = admittance × *concentration* in the flowing phase, then, for both heat or mass into the jth phase,

$$\partial \bar{v}_j(t)/\partial t = \mathbf{Y} v_1(t) = \mathbf{Y} V_1 e^{i\omega t} \tag{7.1a}$$

and further,

$$\bar{v}_j = \bar{V}_j e^{i\omega t}$$

where the bar indicates an instantaneous average over infinitesimal dimension dz; i.e., $\bar{v} = \iint v\,dx\,dy / \iint dx\,dy$;

$$\therefore \quad \partial \bar{v}_j/\partial t = i\omega \bar{V}_j e^{i\omega t} = \mathbf{Y} V_1 e^{i\omega t}$$

i.e.,

$$i\omega \bar{V}_j = \mathbf{Y} V_1 \tag{7.1b}$$

where $\mathbf{Y} = Y_1 + iY_2$ and $v \equiv$ concentration of mass or heat; in the latter case $v \equiv F \times$ temperature. Note that in the above definition *concentration* appears on both sides of the equality, rather than flux on one side and concentration on the other. The dimensions of the shunt admittance thus turn out to be $[T]^{-1}$. Now,

$$q_j = -\tilde{V}_j\,\partial \bar{v}_j/\partial t \tag{7.2}$$

where $\tilde{V}_j$ is the volume of the jth reservoir phase ($j = 2, 3, \ldots$) per unit volume of the flowing phase ($j = 1$). Hence, by Equations (7.1) and (7.2) it follows that

$$q_j = -\tilde{V}_j \mathbf{Y} v_1 = -\tilde{V}_j \mathbf{Y} V_1 e^{i\omega t} \tag{7.3a}$$

Hence,

$$Q_j e^{i\omega t} = -\tilde{V}_j \mathbf{Y} V_1 e^{i\omega t}$$

i.e.,

$$Q_j = -\tilde{V}_j \mathbf{Y} V_1 = -\tilde{V}_j |\mathbf{Y}| V_1 e^{i\phi} \tag{7.3b}$$

or

$$|Q_j| = -\tilde{V}_j |\mathbf{Y}|\,|V_1| \tag{7.3c}$$

where $|\mathbf{Y}| = (Y_1{}^2 + Y_2{}^2)^{1/2}$ and $\phi = \tan^{-1}(Y_2/Y_1)$. Also, from Equations (7.1b) and (7.3b), it follows that

$$i\omega \bar{V}_j = -Q_j/\tilde{V}_j \tag{7.3d}$$

the negative sign being put in only as a reminder that a rising concentration of the reservoir phase means that the direction of the flux is from the flowing phase.

It follows that Equation (1.5a), viz.,

$$D\frac{\partial^2 v}{\partial z^2} - U\frac{\partial v}{\partial z} - \frac{\partial v}{\partial t} + q_j = 0$$

becomes

$$\left[D\frac{\partial^2 V_1}{\partial z^2} - U\frac{\partial V_1}{\partial z} - i\omega V_1 - \tilde{V}_j \mathbf{Y} V_1\right]e^{i\omega t} = 0 \tag{7.3e}$$

Expressions for $\mathbf{Y}$ may be obtained in either of two ways from the analytic expressions for the behavior of the reservoir phase undergoing cyclic variation. One is to find the spatial mean concentration in the reservoir phase and then use equation (7.1a). The other is to find an expression for the flux in the reservoir phase just at the interface between this and the flowing medium, for

$$q_j = \tilde{a}a[\partial v_j(y, t)/\partial y]_{y=0}$$

where y is the direction normal to the flow/reservoir interface.

An example of the step-by-step derivation of a shunt admittance is given now for a simple case (Rosen and Winsche [1]).

Example

(i) *Geometry:* irrelevant.
(ii) *External resistance:* finite, linear, constant; see Section 7.4.2.
(iii) *Internal resistance:* nil (i.e., infinite diffusivity); hence $\bar{v}_j \equiv v_j$.

Figure 7.1 illustrates this case.

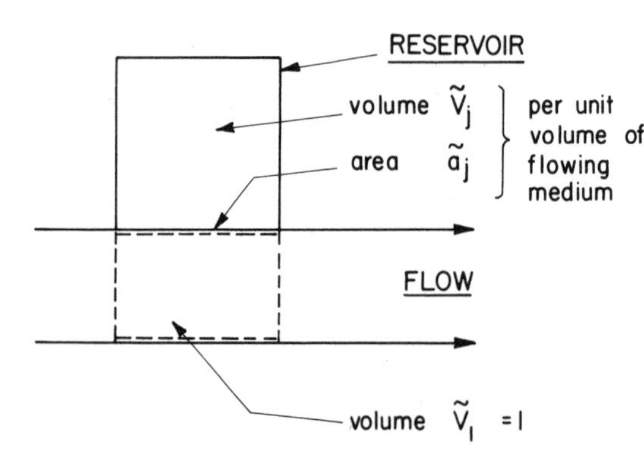

Fig. 7.1. Example of reservoir phase.

Let k_1 denote the transfer coefficient from the flowing phase and k_j denote the transfer coefficient from the reservoir phase. That is, flux/unit area $= k \times$ concentration and the dimensions of k are $[L][T]^{-1}$. The equilibrium constant will have to be incorporated into an expression involving k_1 and k_j. Then, as explained in Appendix 8,

$$-q_j = \tilde{a}_j(k_1 v_1 - k_j \bar{v}_j/\mathbf{K}) = \tilde{V}_j\, \partial\bar{v}_j/\partial t \tag{7.4}$$
$$[= \tilde{a}_j(k_1{}'\mathbf{K}v_1 - k_j{}'\bar{v}_j)]$$

For convenience,

$$\kappa_1 = \tilde{a}_j k_1/\bar{V}_j \qquad \text{or} \qquad \kappa_1' = \tilde{a}_j k_1' \mathbf{K}/\bar{V}_j \tag{7.5}$$

and

$$\kappa_j = \tilde{a}_j k_j/(\bar{V}_j \mathbf{K}) \qquad \text{or} \qquad \kappa_2' = \tilde{a}_j k_j'/\bar{V}_j \tag{7.6}$$

both being of dimensions $[\mathrm{T}]^{-1}$.

Thus, Equation (7.4) becomes

$$(\partial \bar{v}_j/\partial t) + \kappa_j \bar{v}_j = \kappa_1 v_1 \tag{7.7}$$

Since, for the steady cyclic state $v = Ve^{i\omega t}$, it follows that Equation (7.7) becomes

$$\bar{V}_j = [\kappa_1/(i\omega + \kappa_j)]V_1 \tag{7.8}$$

If this is compared with Equation (7.1b) it will be seen that

$$\mathbf{Y} = i\omega\kappa_1/(i\omega + \kappa_j) = Y_1 + iY_2 \tag{7.9}$$

A little manipulation gives

$$Y_1 = \kappa_1/[1 + (\kappa_j/\omega)^2], \quad [\mathrm{T}]^{-1} \tag{7.10}$$

and

$$Y_2 = \kappa_1(\kappa_j/\omega)/[1 + (\kappa_j/\omega)^2], \quad [\mathrm{T}]^{-1} \tag{7.11}$$

Values of Y_1/κ_1 and Y_2/κ_1 are given in Figure 7.2 as functions of ω/κ_2.

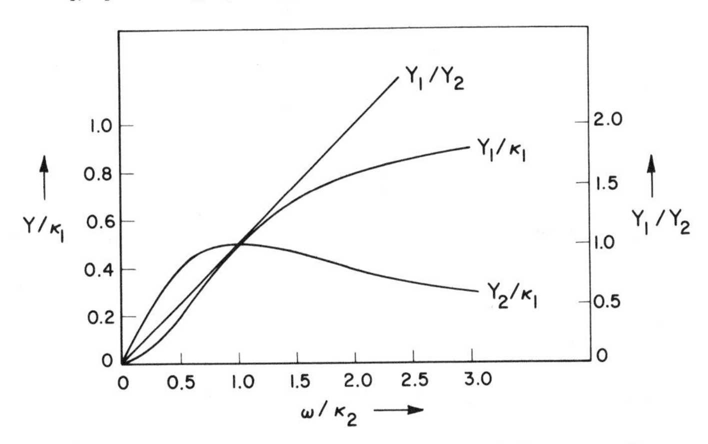

Fig. 7.2. Admittance of system in Figure 7.1. (With acknowledgments to Rosen and Winsche [1].)

7.3 MOMENTS OF A PULSE WAVE

In order that moments of the traveling wave pulse be found, it is necessary that an expression for q in the Laplace domain be derived. That is to say, the partial differential equation and boundary conditions of the reservoir phase must be set up and transformed to distance and Laplace variables.

Example (Chao and Hoelscher [2])

The reservoir phase is as described in the previous section. The differential equation is

$$\partial \bar{v}_j(t)/\partial t = \kappa_1 v_1(t) - \kappa_j \bar{v}_j(t) \tag{7.7) R}$$

Its Laplace transformation gives

$$\bar{\bar{v}}_j(s) = \kappa_1 \bar{v}_1(s)/(s + \kappa_j) \tag{7.12}$$

Also,

$$q(t) = -\tilde{V}_j\, \partial \bar{v}_j(t)/\partial t \tag{7.2) R}$$

so,

$$\bar{q}(s) = -\tilde{V}_j s \bar{\bar{v}}_j(s) = -\tilde{V}_j s \kappa_1 \bar{v}_1(s)/(s + \kappa_j) \tag{7.13}$$

by Equation (7.12), a result which could have been obtained directly by changing $i\omega$ to s in the previous example.

7.4 THE SHUNT ADMITTANCE OF SOME IMPORTANT CASES

Spheres and prisms (no flux at any surface except interface) will be dealt with. These are useful models, and the results are mathematically related; see Carslaw and Jaeger [3]. Gunn [4] discusses other cases. In what follows, the subscript j will usually be omitted from both a and $\tilde{a}$.

7.4.1 No External Resistance (between Flowing Phase and Reservoir); Finite Diffusivity

7.4.1.1 Geometry: Sphere, Radius r_0, Homogeneous, Isotropic

Rosen and Winsche [1], from Carslaw and Jaeger [3, p. 235], give

$$Y_1 = (3\mathbf{K}a/r_0^2)H_1(r_0\omega'), \quad [\mathrm{T}]^{-1} \tag{7.14}$$

$$Y_2 = (3\mathbf{K}a/r_0^2)H_2(r_0\omega'), \quad [\mathrm{T}]^{-1} \tag{7.15}$$

where

$$H_1 = \{r_0\omega'[\sinh(2r_0\omega') + \sin(2r_0\omega')]/[\cosh(2r_0\omega') - \cos(2r_0\omega')]\} - 1 \tag{7.16}$$

$$H_2 = r_0\omega'[\sinh(2r_0\omega') - \sin(2r_0\omega')]/[\cosh(2r_0\omega') - \cos(2r_0\omega')] \tag{7.17}$$

$$\omega' = (\omega/2a)^{1/2} \tag{7.18}$$

Values of H_1 and H_2 are given as functions of $r_0\omega'$ in Figure 7.3.

7.4.1.2 Geometry: Prism, Length l, Homogeneous, Isotropic

Figure 7.4 illustrates this case. (It is assumed that the dimension of the exposed face f in the direction of flow is small compared with the length of

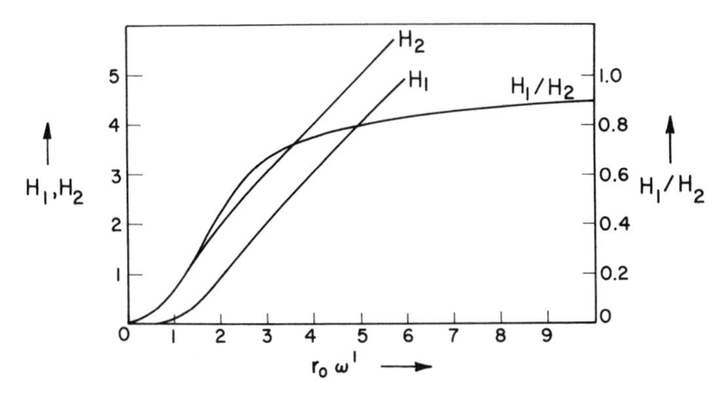

Fig. 7.3. Functions H_1 and H_2 (Case 7.4.1.1). (With acknowledgments to Rosen and Winsche [1].)

a wave.) Turner [5], from Carslaw and Jaeger [3, p. 105], gives

$$Y_1/\omega = (\mathbf{K}/2l\omega') \times [\sinh(2l\omega') - \sin(2l\omega')]/[\cos(2l\omega') + \cosh(2l\omega')], \quad [0] \tag{7.19}$$

$$Y_2/\omega = (\mathbf{K}/2l\omega') \times [\sinh(2l\omega') + \sin(2l\omega')]/[\cos(2l\omega') + \cosh(2l\omega')], \quad [0] \tag{7.20}$$

Values of Y_1/ω and Y_2/ω are given in Figure 7.5; both tend to the value $1/(2l\omega')$ when $2l\omega' > 3$. Again when $2l\omega' \to 0$, then $Y_1/\omega \to (2l\omega')^2/6$ and $Y_2/\omega \to 1$.

7.4.1.3 Prism As in Case 7.4.1.2 but $l \to \infty$

The difficulty (not insuperable) caused by taking l to the limit in Equations (7.19) and (7.20), and by the fact that $\tilde{V}_j \to \infty$, is obviated by making use of the expression for the flux per unit area of surface f (Figure 7.4) in Carslaw and Jaeger [3, p. 67]. There results

$$(\tilde{V}_j Y_1) = (\tilde{V}_j Y_2) = \mathbf{K}\tilde{a}_j(\omega/2\omega') \tag{7.21}$$

i.e., from Equation (7.3a),

$$q_j = -[\mathbf{K}\tilde{a}\omega/2\omega'][1 + i]v_1 \tag{7.22}$$

where q_j is in, for example, mal sec^{-1} cm^{-3}, i.e., per unit volume of flowing medium. ("Mal" means mol or cal, as appropriate.)

From Equation (7.22) it will be seen that the flux wave leads the concentration wave in the flowing fluid by $\pi/4$ radians. This is not true for Case 7.4.1.2, where the phase angle $[\equiv \tan^{-1}(Y_2/Y_1)] = f(l) \neq \pi/4$. These contrasting results may be thought of as showing the presence and absence of reflections from the far wall in the two cases; see Chapter 6.

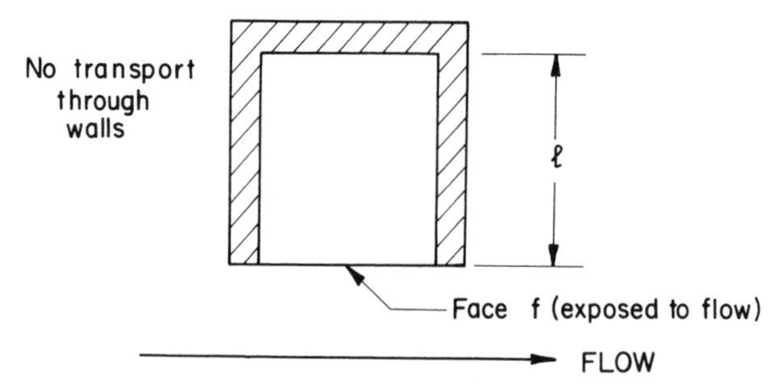

Fig. 7.4. Reservoir of Case 7.4.1.2.

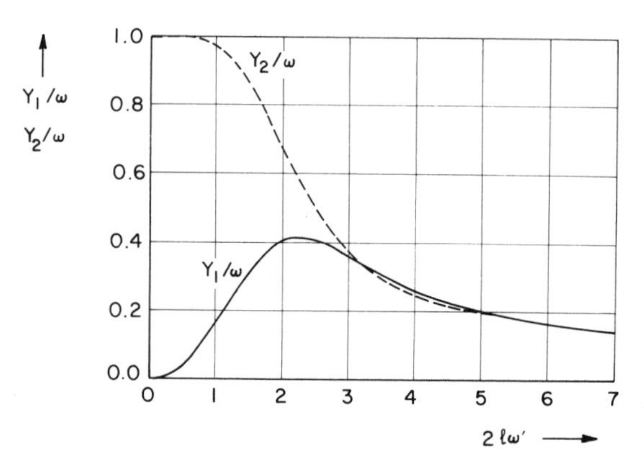

Fig. 7.5. Admittance of system in Figure 7.4 (**K** = 1).

7.4.2 Finite External Resistance

If the resistance resides in a boundary layer of small but finite mass, as it usually does, it has its own response, i.e., it has a complex impedance of its own, but the practice is to assume that it acts as a pure resistance and thus causes no phase-angle change across it. That is, if the reservoir phase without external resistance has shunt admittance $\mathbf{Y} = 1/\mathbf{Z}$, then when a series resistance R (a scalar) does exist the admittance is changed to $\mathbf{Y}' = 1/(\mathbf{Z} + R)$, and R must have the same units and dimensions as $\mathbf{Z}$, viz., of time. In general (Appendix 8),

$$R = \tilde{V}_j/\tilde{a}_j k, \quad [\mathrm{T}]$$

where

$$k = \text{mass transfer coefficient} \tag{7.23}$$

or

$$k = (1/F_1) \times (\text{heat transfer coefficient}) \tag{7.24}$$

i.e., both are in units of, for example, mal sec^{-1} cm^{-2} mal^{-1} cm^3 $\equiv$ [L][T]$^{-1}$.
It follows that

$$\mathbf{Y}' = (R[Y_1^2 + Y_2^2] + Y_1 + iY_2)/(R^2[Y_1^2 + Y_2^2] + 2RY_1 + 1) \qquad (7.25)$$

Alternatively, the appropriate differential equation and boundary condition may be solved; some examples of the results follow.

7.4.2.1 (a) Solid Sphere (finite diffusivity), Radius r_0

Here,

$$R = r_0/3k \qquad (7.26)$$

generally and

$$R = r_0 F_1/3h \qquad (7.27)$$

for heat transfer. The results for this case may be found either by the introduction of resistance to the case of a sphere without external resistance (viz., Case 7.4.1.1) or from the appropriate differential equations and conditions. These and a solution are given in Carslaw and Jaeger [3, p. 238]. The derivation is done in full in Goss [6] and the results are given in another form in Turner [7]. They are

$$Y_1 = (1/R)\Lambda \qquad (7.28)$$

$$Y_2 = (1/R)\Xi \qquad (7.29)$$

where

$$\Lambda = A\cos\phi_0 - B\sin\phi_0 = \Re[(A + iB)\exp i\phi_0] \qquad (7.30)$$

$$\Xi = A\sin\phi_0 + B\cos\phi_0 = \Im[(A + iB)\exp i\phi_0] \qquad (7.31)$$

where

$$A = (n\Sigma - \zeta)/[(2\zeta)^{1/2} a_2] \qquad (7.32)$$

$$B = n\sigma/[(2\zeta)^{1/2} a_2] \qquad (7.33)$$

$$\Sigma = \sinh 2n + \sin 2n \qquad (7.34)$$

$$\zeta = \cosh 2n - \cos 2n \qquad (7.35)$$

$$\sigma = \sinh 2n - \sin 2n \qquad (7.36)$$

$$a_2 = n\{\Gamma + (N - 1)(2\zeta/2n^2) + [\Sigma(N - 1)/n]\}^{1/2} \qquad (7.37)$$

$$N = \text{Biot number} = r_0 k/a_j\mathbf{K} \qquad (7.38)$$

$$\Gamma = \cosh 2n + \cos 2n \qquad (7.39)$$

$$m = r\omega' \qquad (7.40)$$

$$n = r_0\omega' \qquad (7.41)$$

$$\phi_0 = \phi_1(r_0) - \phi_2 \tag{7.42}$$

$$\phi_1(r) = \tan^{-1}(\coth m \tan m) \tag{7.43}$$

$$\phi_2 = \tan^{-1}\frac{1 + \tanh n \tan n + (N - 1)[\tan n)/n]}{1 - \tanh n \tan n + (N - 1)[(\tanh n)/n]} \tag{7.44}$$

A program exists (Goss [6]) to calculate the above values on an IBM 360.75 computer.

7.4.2.1(*b*) *Hollow Sphere*

The shunt admittance has been derived in Otten [8].

7.4.2.2 *Finite Prism with Resistance*

Figure 7.4 shows the prism of length l. Either the result of Case 7.4.1.2 may be used by adding a resistance term, or Equations (5)–(7) in §3.12 of Carslaw and Jaeger [3] can be made to give an expression for the shunt admittance. It is

$$\mathbf{Y} = -(\omega' k/l)([(A_0'C_l + B_0'D_l) - i(A_0'D_l - B_0'C_l)]/[C_l^2 + D_l^2]) \tag{7.45}$$

where

$$A(x) = \cosh x\omega' \cos x\omega' \tag{7.46}$$

$$B(x) = \sinh x\omega' \sin x\omega' \tag{7.47}$$

$$C_l = \omega' \sinh l\omega' \cos l\omega' - \omega' \cosh l\omega' \sin l\omega' + (k/a\mathbf{K}) \cosh l\omega' \cos l\omega' \tag{7.48}$$

$$D_l = \omega' \sinh l\omega' \cos l\omega' + \omega' \cosh l\omega' \sin l\omega' + (k/a\mathbf{K}) \sinh l\omega' \sin l\omega' \tag{7.49}$$

and A_0' (etc.) means $[dA/d(x\omega')]_{x=0}$ (etc.).

7.4.2.3 *Infinite Prism with Resistance*

The result may be deduced from the result of Case 7.4.1.3 (infinite prism without resistance) by adding the effect of resistance as mentioned in Section 7.4.2. Alternatively, Equation (4) of §2.8 in Carslaw and Jaeger [3] can be used to give the flux at the interface in terms of the concentration in the flowing phase and Y. There arises

$$\tilde{V}_j Y_1 = \frac{\tilde{a}k\omega'[k/(a\mathbf{K}) + 2\omega']}{\text{Denominator}} \tag{7.50}$$

$$\tilde{V}_j Y_2 = \frac{\tilde{a}k\omega'[k/a\mathbf{K}]}{\text{Denominator}} \tag{7.51}$$

where

$$\text{Denominator} \equiv [k/(a\mathbf{K}) + \omega']^2 + (\omega')^2$$

When $h \longrightarrow \infty$ these expressions revert to those for Case 7.4.1.3, as they should. Note that in the present case the absence of a reference length prevents the use of the dimensionless groups N and (length $\times$ ω') which occurred in Case 7.4.2.2.

7.4.3 Summary (Sine Waves)

The equations in Chapter 6 give the amplitude and phase angle in the *flowing stream relative to those at some reference location upstream.* They require a knowledge of the flux entering the flow stream from the reservoir.

The equations in the first part of this chapter give, by means of the admittance, the amplitude and phase angle of *the flux entering the flow stream at any point relative to the temperature or concentration in the stream* at that point. The relative amplitude is, by Equation (7.3c), equal to $\tilde{V}_j\,[Y_1{}^2 + Y_2{}^2]^{1/2}$ and the phase angle $= \tan^{-1}(Y_2/Y_1)$.

The propagation of the wave in the reservoir may be found from the appropriate differential equation; see, for example, Carslaw and Jaeger [3], Moore [9], Schneider [10, 11], Mikheyev [12], and Jakob [13].

7.4.3.1 The Phase Angle between the Concentration in the Flowing Phase and Flux into the Reservoir

This phase angle will depend, of course, upon the attributes of the reservoir phase; it will also depend upon the magnitude of the resistance to transfer between the phases. As regards the latter effect, it will be seen from Equation (7.25) that no matter what the values and sign of Y_1 and Y_2 are, the phase angle, viz.,

$$\phi = Y_2/(R[Y_1{}^2 + Y_2{}^2] + Y_1) \tag{7.52}$$

will be found to decrease as R increases. The resistance R is always positive, and in the situations discussed here Y_1 and Y_2 are also positive. Even if they were not, the absolute magnitude of the (negative) phase angle would decrease. Thus, if the equations for Y_1 and Y_2 listed above be examined, it will be found that the phasor of flux, relative to the phasor of concentration, lies in the first quadrant; i.e., the phase angle can range from $\pi/2$ to zero. As $R \longrightarrow \infty$, the (vanishingly small) flux tends to be in phase with the wave in the flowing phase, which seems logically acceptable. Some cases are illustrated in Figure 7.6.

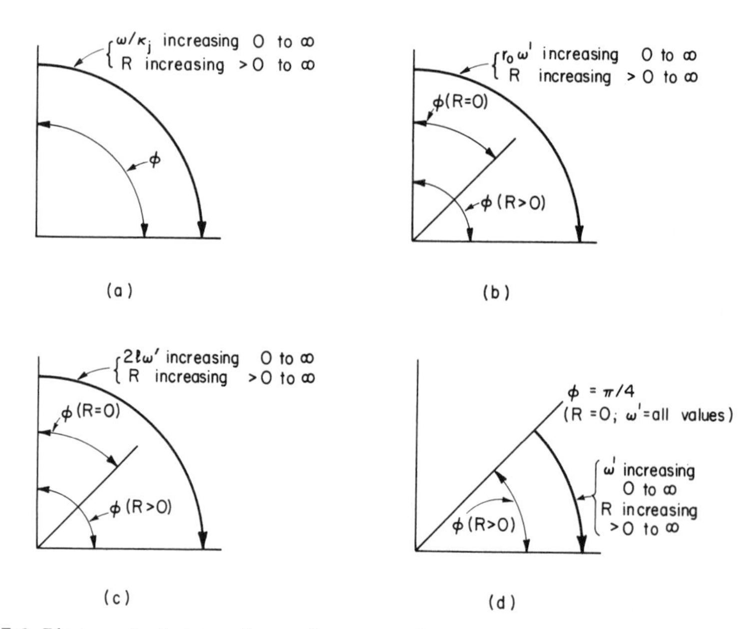

Fig. 7.6. Phase angles between flux and concentration vectors at the surface of the following reservoirs: (a) No internal resistance ($R > 0$). (b) Sphere. (c) Finite slab. (d) Semiinfinite slab.

7.5 RESERVOIRS IN SERIES

Composite solids, and composite processes involving transfer from phase to phase, occur in both heat and mass transfer processes. In heat transfer the behavior of composite solids, such as, for example, layers of slabs as in Figure 4.3, or concentric spherical or cylindrical shells, is discussed in Carslaw and Jaeger [3] and references therein; the analogy with electrical circuits is there brought out. Another class of composite solid comprises grains of different solids cemented together, or of grains of solid interspersed with empty pores; ore or fertilizer pellets and thermal insulation are important industrial examples. While theory is developing aimed at allowing the gross conductivity to be computed from the values of the conductivities of the individual solids, from the point of view both of the use of waves and of industrial design, determination of the overall thermal conductivity is satisfactory. It allows of engineering calculations, while reducing the number of parameters (see Chapter 9) to be found.

On the other hand, in mass transport there are often many disparate processes; not only is an "overall" transport coefficient unsatisfactory in the design process, but also the purpose behind using kinematic waves is very often that of investigating the relative magnitudes of the separate processes. Chromatography, catalysis, and plant nutrient storage are three important examples in which material moves from the flowing phase in processes that are in series. Since the ultimate repository of the matter is the surface of the solid, with but little or no penetration, it follows that the provision of sufficient surface per unit volume of flow, without the embarrassments of small channels and large pressure drop, requires a solid with a large amount of internal surface. (*N.B.* Concentrations "in" the solid may still be expressed as amount per unit volume or mass; the equilibrium constant is defined accordingly.) This surface in turn demands access via pores, and these need to have their geometry specified in order that a differential equation may be written.

The model of Kučera [14] is illustrative and important because others put forward earlier (see Chapter 9) were special cases. It is shown in Figure 7.7.

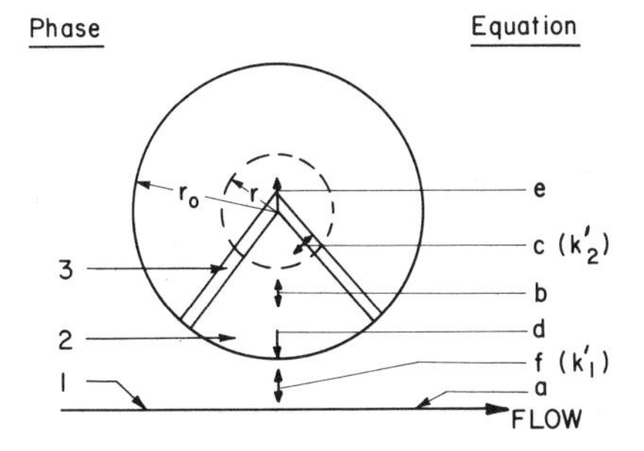

Fig. 7.7. Kučera's [14] model of chromatographic system.

Kučera's model assumes a sphere of nonflowing phase in which mass transport is governed by the differential equation (7.53b). That is, the surfaces of equal concentration at any time are concentric spheres as shown dashed. However, the volume of this nonflowing fluid is only a fraction of the sphere's volume, and the area of the mouth of the pore communicating with the flowing fluid (evenly all around) is only a fraction of the surface area, and surfaces in the solid provide another reservoir "in series" with the radial diffusional process in the pore. (The differential equation for diffusion in the pore is unaltered, apart from the addition of another term.)

Thus, the equations are as follows.

In the flowing fluid (*Phase 1*):

$$D\frac{\partial^2 v_1}{\partial z^2} - U\frac{\partial v_1}{\partial z} - \frac{\partial v_1}{\partial t} + q_1 = 0 \tag{7.53a}$$

where q_1 is the flux from the pore.

In the "spherical pore" (Phase 2):

$$a_2\left(\frac{\partial^2 v_2}{\partial r^2} + \frac{2}{r}\frac{\partial v_2}{\partial r}\right) - \frac{\partial v_2}{\partial t} + q_2 = 0 \tag{7.53b}$$

where a_2 is the diffusion coefficient in the pore and q_2 is the flux from the adsorbing surface.

At the "pore to adsorbing surface" interface (adsorbing surface is Phase 3):

$$q_2 = -k_2'(\mathbf{K}_2 v_2 - v_3) \tag{7.53c}$$

(v_3 is the concentration on the surface expressed as, for example, moles per unit volume of pores.)

At the mouth of the pore:

$$q_1 = \frac{\epsilon_{\text{int}}}{\epsilon_{\text{ext}}}(1 - \epsilon_{\text{ext}})\frac{3a}{r_0}\left(\frac{\partial v_2}{\partial r}\right)_{r=r_0} \tag{7.53d}$$

where ϵ_{int} is the fractional volume of pore in a sphere, ϵ_{ext} is the fractional volume of spheres in a bed, and $3/r_0$ gives the area of the sphere's surface per unit volume of sphere.

At the center of the pore:

$$(\partial v_2/\partial r)_{r=0} = 0 \tag{7.53e}$$

Between mouth of pore and flowing fluid:

$$q_1 = -k_1'[\mathbf{K}_1 v_1 - v_2(r_0)] \tag{7.53f}$$

In the above k_1' and k_2' are mass transfer coefficients which will incorporate relative areas, eventually related to unit volume of flowing medium, depending on the model and assumptions.

The intent is to eliminate all concentrations except one—usually that in the flowing stream, viz., v_1—but it is first necessary to specify the initial conditions. Usually these specify a uniform concentration in the solid phase with a change in the concentration of the flowing phase; for example, it may suffer an impulse or a step change at $t = 0$, or a finite pulse from $t = 0$ to $t = t_0$, or a generalized pulse $f(t)$, or a steady, sinusoidal variation. For the first three cases an analytic solution would be difficult or impossible, but if the Laplace transformations of these equations be taken before elimination of the ancillary concentrations, then differential equations in z and r are obtained. These are then solved, and all concentrations except one (e.g., v_1), eliminated. The result is an expression for $\bar{v}_1(z, s)$ as a function of z and the Laplace variable s. Inversion is difficult or impossible, but either the steady cyclic state is considered by putting $s = i\omega$, or expressions for the moments are found by the process described in Chapter 4. Thus, for a Dirac input into a system with spherical packing initially all at zero concentration

Kučera gives an equation for $\bar{v}_1(L, s)$ and expressions for the zeroth, first, . . . , fifth moments, the higher moments growing rapidly in complexity (the expression for the fifth moment comprises 19 lines of algebra). These results, and those of other authors, refer to an infinite bed, i.e., no boundary conditions are inserted into the bed. In principle, these boundary conditions could be specified, but it was seen in Chapter 5 how the analysis becomes rapidly more complicated even when there was no reservoir phase; when there is such a reservoir system, even of the simplest kind, the workings would probably be impossibly tedious.

7.6 NUMERICAL EXAMPLES (SINE WAVES)

A few specimen numerical calculations of admittance are given now. They illustrate case of:

1. Mass transfer. Finite external, zero internal resistance (geometry is not required to be specified).
2. Mass transfer. Zero external, finite internal resistance. Geometry is a prism.
3. As Example 2, but the phase is infinite normal to the flow.
4. Heat transfer. Finite external, finite internal resistance. Geometry is a sphere.

The negative sign of Equation (7.3) will be dropped.

Problem 1 (Mass transfer)

A reservoir phase is of infinite diffusivity, and offers an interphase area $\tilde{a}_2$ of 10 cm²/cm³ of flowing phase, and a volume $\tilde{V}_2$ of 100 cm³/cm³ of flowing phase.

The mass transfer rate constants: $k_1' = k_2' = 8 \times 10^{-2}$ cm sec^{-1}.
Frequency: $\omega = 1.2 \times 10^{-2}$ rad sec^{-1}.
Equilibrium constant: $\mathbf{K} = 0.65$.

Find the phase lag and amplitude of the mass flux into the reservoir phase (relative to the sine wave in the flowing phase).

Answer

From Equations (7.5) and (7.6), viz.,

$$\kappa_1' = \tilde{a}_2 \mathbf{K} k_1'/\tilde{V}_2, \qquad \kappa_2' = \tilde{a}_2 k_2'/\tilde{V}_2$$

$$\kappa_1' = (10 \times 0.65 \times 8 \times 10^{-2})/100 \times 1 = 5.2 \times 10^{-3} \quad \text{sec}^{-1}$$

$$\kappa_2' = 10 \times 8 \times 10^{-2}/100 \times 1 = 8.0 \times 10^{-3} \quad \text{sec}^{-1}$$

Hence, $\omega/\kappa_2' = 12 \times 10^{-3}/8.0 \times 10^{-3} = 1.5$. Calculation from Equations

(7.10) and (7.11) (or the use of Figure 7.2) gives

$$Y_1/\kappa_1' = 0.7 \qquad \text{and} \qquad Y_2/\kappa_1' = 0.46$$

Hence,

$$Y_1 = 0.7 \times 5.2 \times 10^{-3} = 3.64 \times 10^{-3} \quad \text{sec}^{-1}$$
$$Y_2 = 0.46 \times 5.2 \times 10^{-3} = 2.392 \times 10^{-3} \quad \text{sec}^{-1}$$

Thus, by Equation (7.3b)

$$Q = \tilde{V}_2(Y_1 + iY_2)V_1 \quad \text{(vol/vol) (time}^{-1}\text{) (concentration)}$$

Hence,

$$Q = \tilde{V}_2 |\mathbf{Y}| V_1 e^{i\phi}$$

where

$$|\mathbf{Y}| = [(3.64 \times 10^{-3})^2 + (2.392 \times 10^{-3})^2]^{1/2} = 4.35 \times 10^{-3} \quad \text{sec}^{-1}$$
$$\tilde{V}_2 |\mathbf{Y}| = 100 \times 4.35 \times 10^{-3} = 4.35 \times 10^{-1} \quad \text{cm}^3\ \text{cm}^{-3}\ \text{sec}^{-1}$$
$$\tan \phi = Y_2/Y_1 = 2.392/3.64 = 0.656$$

or $\phi = 0.58$ rad. Hence,

$$Q = 4.35 \times 10^{-1} V_1 e^{0.58i} \quad \text{mol cm}^{-3}\ \text{sec}^{-1}$$

as in Figure 7.8.

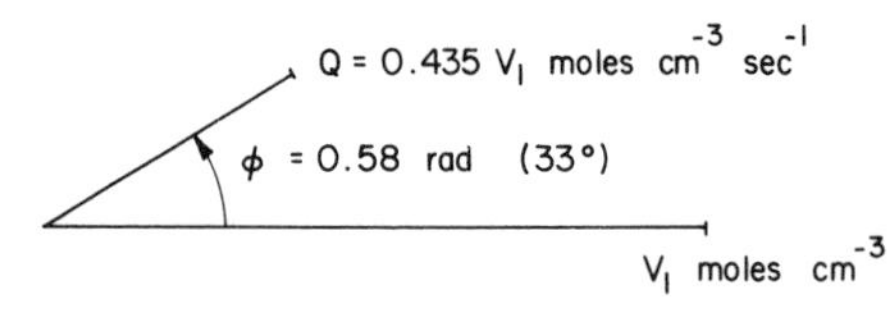

Fig. 7.8. Flux and concentration vectors (Problem 1).

Problem 2 (Mass transfer)

A reservoir phase has the following properties:

Geometry: prism of length $l = 0.2$ cm, area $\tilde{a} = 3.75$ cm^2 cm^{-3}, volume $\tilde{V} = l\tilde{a} = 0.75$ cm^3 cm^{-3}.

Resistances: external, nil; *internal,* finite; diffusion coefficient $a = 4 \times 10^{-6}$ cm^2 sec^{-1}, equilibrium constant $\mathbf{K} = 1.0$.

Frequency: $\omega = 2 \times 10^{-4}$ rad sec^{-1}

Find the flux into the reservoir relative to the concentration in the flowing phase.

Answer

By definition

$$\omega' = (\omega/2a)^{1/2} = (2 \times 10^{-4}/2 \times 4 \times 10^{-6})^{1/2} = 5.0 \quad \text{cm}^{-1}$$

Hence, $2\omega' l = 2 \times 5.0 \times 0.2 = 2.0$.

From Figure 7.5 or by calculation from Equations (7.19) and (7.20) it is found that

$$Y_1/\omega = 0.41, \qquad Y_2/\omega = 0.68$$

i.e.,

$$Y_1 = 2 \times 10^{-4} \times 0.41 = 8.2 \times 10^{-5} \quad \text{sec}^{-1}$$

$$Y_2 = 2 \times 10^{-4} \times 0.68 = 13.6 \times 10^{-5} \quad \text{sec}^{-1}$$

Hence, flux $Q = \tilde{V}_2 |\mathbf{Y}| Y_1 e^{i\phi}$, and since

$$|\mathbf{Y}| = [(8.2 \times 10^{-5})^2 + (13.6 \times 10^{-5})^2]^{1/2} = 15.75 \times 10^{-5} \quad \text{sec}^{-1}$$

so

$$\tilde{V}_2 |\mathbf{Y}| = 0.75 \times 15.75 \times 10^{-5} = 11.88 \times 10^{-5} \quad \text{cm}^3\ \text{cm}^{-3}\ \text{sec}^{-1}$$

and

$$\phi = \tan^{-1}(13.6/8.2) = \tan^{-1} 1.66 = 1.03 \quad \text{rad} \quad (59°)$$

Hence

$$Q = 11.88 \times 10^{-5} V_1 e^{1.03i} \quad \text{mol cm}^{-3}\ \text{sec}^{-1}$$

as in Figure 7.9.

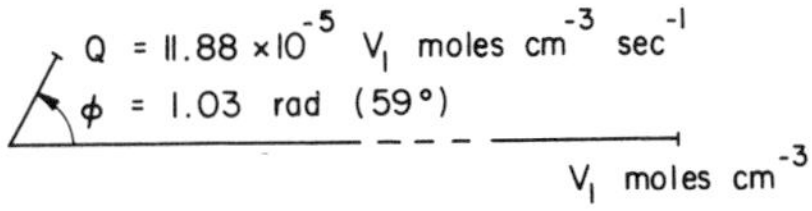

Q = 11.88 × 10^{-5} V$_1$ moles cm^{-3} sec^{-1}
φ = 1.03 rad (59°)
V$_1$ moles cm^{-3}

Fig. 7.9. Flux and concentration vectors (Problem 2).

Problem 3 (Mass Transfer)

As Problem 2 but $l \rightarrow \infty$, and $\tilde{a} = 1.5$ cm^2 cm^{-3}.

Answer

As in Example 2, $\omega' = 5.0$.
By Equation (7.22),

$$Q = [\mathbf{K}\tilde{a}\omega/2\omega'](1 + i)V_1$$
$$= [(1 \times 1.5 \times 2 \times 10^{-4})/(2 \times 5.0)](1 + i)V_1$$

and

$$(1 + i) \equiv \sqrt{2}\, e^{i\pi/4}$$

So

$$Q = 4.25 \times 10^{-5} V_1 e^{i\pi/4} \quad \text{mol cm}^{-3}\ \text{sec}^{-1}$$

as in Figure 7.10.

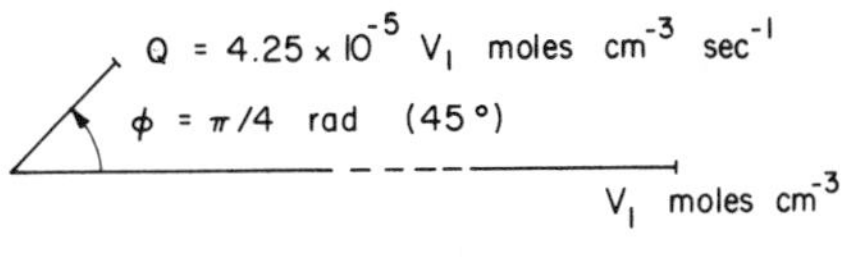

Q = 4.25 × 10^{-5} V$_1$ moles cm^{-3} sec^{-1}
φ = π/4 rad (45°)
V$_1$ moles cm^{-3}

Fig. 7.10. Flux and concentration vectors (Problem 3).

Problem 4 (Heat transfer)

A reservoir phase has the following properties:

Geometry: sphere, of radius $r_0 = 0.2$ cm.
Resistances: External, finite; heat transfer coefficient $h = 25.0 \times 10^{-3}$ cal sec^{-1} cm^{-2} deg^{-1}; $F_1 = 3.1 \times 10^{-4}$ cal cm^{-3} deg^{-1}; $k_1 = k_2 = (25.0 \times 10^{-3}/3.1 \times 10^{-4})$ cm sec^{-1}, i.e., $k_1 = k_2 = 80.6$ cm sec^{-1}. *Internal,* finite; conductivity $\mathbf{k} = 2.5 \times 10^{-3}$ cal sec^{-1} deg^{-1} cm^{-1}; $F_2 = 0.5$ cal cm^{-3} deg^{-1}; i.e., $a = 2.5 \times 10^{-3}/0.5 = 5.0 \times 10^{-3}$ cm^2 sec^{-1}.
Frequency: $\omega = 2.5 \times 10^{-3}$ rad sec^{-1}.
Bed porosity: $\epsilon = 0.6$.

Find the flux into the reservoir relative to the concentration in the flowing phase.

Answer

$r_0\omega'$ is small (<1). See note below.

[*Note*

In general the answer requires that computations be done using Equations (7.26) through (7.44). These are straightforward but tedious; as noted earlier, a computer program exists, listed in Goss [6], designed to compute the response of a flowing medium with reservoir phase. This program, containing the quantities Λ and Ξ, could be amended to print out their values, for Equations (7.28) and (7.29) state that

$$Y_1 = (1/R)\Lambda, \qquad Y_2 = (1/R)\Xi$$

The terms Λ and Ξ can be expressed as functions of the dimensionless groups $r_0\omega'$ and $r_0(k/a)(F_1/F_j)$ (the Biot number N). When $r_0\omega'$ is large, then Λ and Ξ have asymptotic values, while when $r_0\omega'$ is small (say, less than 1.0) then they are describable by simple algebraic expressions, as stated by Turner [15]. This is true for all values of the Biot number. The derivation is given in full in Goss [6].

Since

$$r_0\omega' = r_0(\omega/2a)^{1/2} = 0.2(2.5 \times 10^{-3}/2 \times 5 \times 10^{-3})^{1/2} = 0.1$$

the use of these limiting expressions will be demonstrated here.]

The value of the Biot number $N = r_0 h/\mathbf{k} = 0.2 \times 25.0 \times 10^{-3}/2.5 \times 10^{-3} = 2.0$.

Now, by Equations (7.28) and (7.29)

$$Y_1 = (1/R)\Lambda, \qquad Y_2 = (1/R)\Xi$$

where, by Equations (7.30) and (7.31),

$$\Lambda = A\cos\phi_0 - B\cos\phi_0, \qquad \Xi = A\sin\phi_0 + B\cos\phi_0$$

in which, in the limit as $r_0\omega' \to 0\ (r_0\omega' \ll 1)$

$$A \longrightarrow (4/45)(r_0\omega')^4/N$$

$$B \longrightarrow \tfrac{2}{3}(r_0\omega')^2/N$$

$$\phi_0 \longrightarrow -\tfrac{2}{3}(r_0\omega')^2/N$$

Hence,

$$\lim(\Lambda)_{r_0\omega'\to 0} = [4(r_0\omega')^4/9N][0.2 + (1/N)]$$

$$\lim(\Xi)_{r_0\omega'\to 0} = 2(r_0\omega')^2/3N$$

(The value of N, of course, determines what the error in approximating will be.) Hence, since $r_0\omega' = 0.1$, $N = 2.0$, then

$$\Lambda = (4 \times 1 \times 10^{-4}/9 \times 2)[0.2 + (1/2)]$$

i.e., $\Lambda = 1.55 \times 10^{-5}$;

$$\Xi = 2 \times 1 \times 10^{-2}/3 \times 2$$

i.e., $\Xi = 3.33 \times 10^{-3}$; also

$$R = r_0F_1/3h = 0.2 \times 3.1 \times 10^{-4}/3 \times 25.0 \times 10^{-3}$$

i.e., $R = 0.825 \times 10^{-3}\ \text{sec}^{-1}$.

Hence,

$$Y_1 = (1/0.825 \times 10^{-3}) \times 1.55 \times 10^{-5} = 1.89 \times 10^{-2} \quad \text{sec}^{-1}$$

$$Y_2 = (1/0.825 \times 10^{-3}) \times 3.33 \times 10^{-3} = 4.04 \quad \text{sec}^{-1}$$

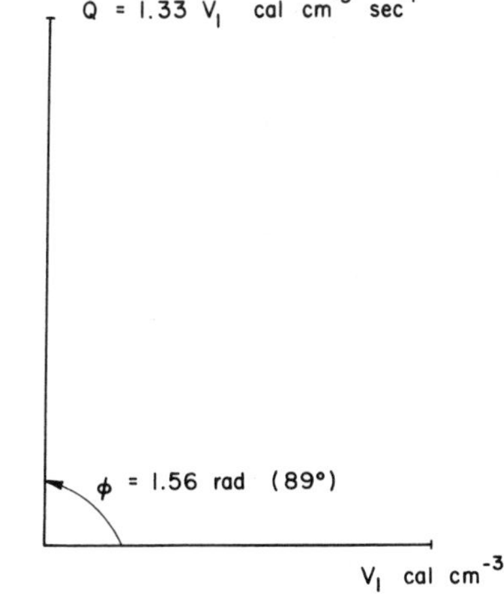

Fig. 7.11. Flux and concentration vectors (Problem 4).

Further, the volume $\tilde{V}$ of reservoir per unit volume of flow $= (1 - \epsilon)/\epsilon = 0.4/0.6 = 0.33$. So, $Q = \tilde{V}|\mathbf{Y}|V_1 e^{i\phi}$ and

$$|\mathbf{Y}| = [(1.89 \times 10^{-2})^2 + (4.04)^2]^{1/2} \approx 4.04 \quad \text{sec}^{-1}$$

$$\tilde{V}|\mathbf{Y}| = 4.04 \times 0.33 = 1.33 \quad \text{sec}^{-1}$$

$$\phi = \tan^{-1}(4.04/1.89 \times 10^{-2}) = \tan^{-1} 213 \approx 1.56 \quad \text{rad} \quad (89°)$$

Hence,

$$Q = 1.33 V_1 e^{1.56i} \quad \text{cal cm}^{-3}\ \text{sec}^{-1}$$

as in Figure 7.11.

REFERENCES

[1] J. B. Rosen and W. E. Winsche, The Admittance Concept in the Kinetics of Chromatography. *J. Chem. Phys.* **18**, 1587 (1950).
[2] R. Chao and H. E. Hoelscher, Simultaneous Axial Dispersion and Adsorption in a Packed Bed. *AIChE J.* **12**, 271 (1966).
[3] H. S. Carslaw and J. C. Jaeger, "Conduction of Heat in Solids." Oxford Univ. Press, London and New York, 1959.
[4] D. J. Gunn, The Transient and Frequency Response of Particles and Beds of Particles. *Chem. Eng. Sci.* **25**, 53 (1970).
[5] G. A. Turner, The Flow-structure in Packed Beds. *Chem. Eng. Sci.* **7**, 156 (1958).
[6] M. J. Goss, Determination of Thermal Parameters of a Packed Bed of Spheres by a Method of High Precision Frequency Response. Ph.D. Thesis, Univ. of Waterloo (1969).
[7] G. A. Turner, A Method of Finding Simultaneously the Values of the Heat Transfer Coefficient, the Dispersion Coefficient, and the Thermal Conductivity of the Packing in a Packed Bed of Spheres: Part 1. Mathematical Analysis. *AIChE J.* **13**, 678 (1967).
[8] L. Otten, Determination of the Thermal Properties of Packed Beds of Small Heterogeneous Particles. Ph.D. Thesis, Univ. of Waterloo (1971).
[9] R. K. Moore, "Traveling Wave Engineering." McGraw-Hill, New York, 1960.
[10] P. J. Schneider, "Conduction Heat Transfer." Addison-Wesley, Reading, Massachusetts, 1955.
[11] P. J. Schneider, "Temperature Response Charts." Wiley, New York, 1963.
[12] M. Mikheyev, "Heat Transfer." Mir Publ., Moscow, 1968.
[13] M. Jakob, "Heat Transfer." Wiley, New York, 1949.
[14] E. Kučera, Contribution to the Theory of Chromatography. Linear Non-equilibrium Elution Chromatography. *J. Chromatog.* **19**, 237 (1965).
[15] G. A. Turner, The Determination of the Thermal Conductivity of Small Particles. Rep. D2, Fisons Fertilizers Limited, Felixstowe, England (1964).

Chapter 8

Experimental Measurements

8.1 INTRODUCTION

The task of the experimenter is first to set up a mathematical model comprehensive enough to describe a physical situation adequately, simple enough to be simulated in an experiment. The time-varying disturbance must be chosen so that it is practicable and allows the extraction of the required information and the demonstration of the kind and size of errors involved—both computational and experimental.

Pulses and Steps

Both of these changes in concentration or temperature have been widely used because mathematical analysis is often straightforward (see earlier).

But not only is the "perfect" pulse or step probably impossible to achieve, but also the essay into producing it leads to a system different from the one defined by the mathematical model. For it is clear that in most systems where mass or thermal inertia is an intrinsic property the "instantaneous rise" of either pulse or step is impossible, and in the case of the pulse (to compound the problem) the requirement of "infinite intensity over infinitesimal time" cannot be satisfied. So the pulse is bounded, both are rounded, and both need a finite time to mature, as shown in Figure 8.1.

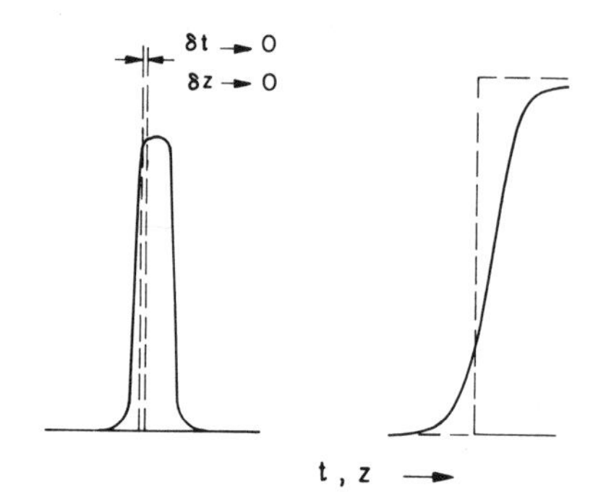

Fig. 8.1. Ideal and actual impulse and step.

(A method that could be used for introducing pulses, steps or sine waves is that of irradiating the flow with an appropriate time-varying beam of electromagnetic radiation. For example, infrared radiation could alter the temperature of a suitably absorbing medium, or visible or ultraviolet light could initiate a fast photochemical reaction; the measurement of the concentration of products then may have its own problems. For example, if the reaction produces a colored product, the departure from linearity between concentration and light adsorption may be embarrassing; see below.)

Hence, for a situation that can be analyzed it may be necessary to use the moments or the Fourier analysis or the Laplace transform of the pulse. Either the input is assumed to be known, and its moments transform, or harmonic analysis specified, or the input wave has to be measured. In particular, "perfect" impulses and steps are resolvable into sums of periodic waves, as mentioned in Chapter 4. Thus, an imperfect pulse or step as in Figure 8.1 may be considered to be one in which numerous frequencies in these series have been attenuated. Many (and often the higher ones) will have effectively been attenuated out of existence. Thus, a series of frequencies of unknown composition (but which are *assumed* to be known) is being put in simultaneously. The subsequent measurement and analysis are, in effect, of a set of frequencies that does not exist as assumed. This is particularly unfortunate in that the higher frequencies, where the amplitude and phase angle are more difficult to measure accurately, are affected more. Hence, it follows that the practical problems are: (1) to make a device that will inject a pulse or step of a shape that is reasonably approximate to the assumed expansions, (2) to make a detector fast enough to accept all reasonable values of the frequencies, and (3) to make a system that tells the experimenter the *true* shape of the input and output waves, pulses, or steps (in terms of its frequency spectrum or moments). The detector will have its own vagaries and so it should be able to inform what effect it itself has on the readings it gives. The last two requirements are the same as those of sine waves, and these will be considered later. The first requirements, however, introduces other considerations, namely the effect that the introduction of a wave will have upon the flow pattern; it may be that the mathematical model assumed is invalidated by the sudden injection of matter or heat—a Dirac impulse is an explosion—and the more perfect the experimental injection, the more the flow is likely to be time-dependent.

Along with the above considerations, the attractive properties of sine waves listed in Chapter 9 would imply that they are the most useful kind of perturbation in this experimental work. If crude waves and rough measurements are good enough, then this is a relatively easy task; if high precision is required, then the problem rapidly increases in difficulty. Some aspects of this will now be considered.

8.2 GENERAL CONSIDERATIONS ON THE GENERATION AND MEASUREMENT OF PRECISE SINE WAVES

Often measurements of amplitude and phase lag have been taken from physical distances on a record, such as a pen recorder trace, or even a photographed oscilloscope trace. These representations, even if accurate, of a relatively slow wave, even if steady, tend often to mislead in that they may be acceptable to the observer as being a (pure) sine wave; the human eye is tolerant of the presence of appreciable quantities of harmonics. Yet harmonics, unsuspected, can lead to appreciable errors in the measured values of amplitude and phase lag when these are obtained by laying a scale on a trace, for example. This can be seen from Figure 8.2, which contains three waves of the same fundamental frequency; one is pure, the second contains a second harmonic, the third contains this second harmonic but with a different phase angle. The effect on the "amplitude" as measured is obvious, while the "phase angle" of the fundamental ψ ($= \omega t_{\text{lag}}$, ω and t_{lag} being measured separately) is not only not zero (as it should be), but has a value that depends on the place where times are measured. For example, in Figure 8.2

$$2t_{\text{lag}} \neq (t_1)_a + (t_2)_a \neq (t_3)_a + (t_4)_a \neq (t_1)_b + (t_2)_b \neq (t_3)_b + (t_4)_b \neq 0 \tag{8.1}$$

Thus, ψ can be found in this manner only for pure waves and then it does not matter where the two waves are positioned laterally, or where t_{lag} is measured (except that measurements of times of lag from one peak or trough to the next are imprecise, because they cannot be exactly located). Figure 8.3 demonstrates the invariance of t_{lag} with measuring position.

If the trace drifts, then the times have to be measured along a line parallel to lines drawn through troughs and crests. The assumption is that the center line of the wave is straight over this region.

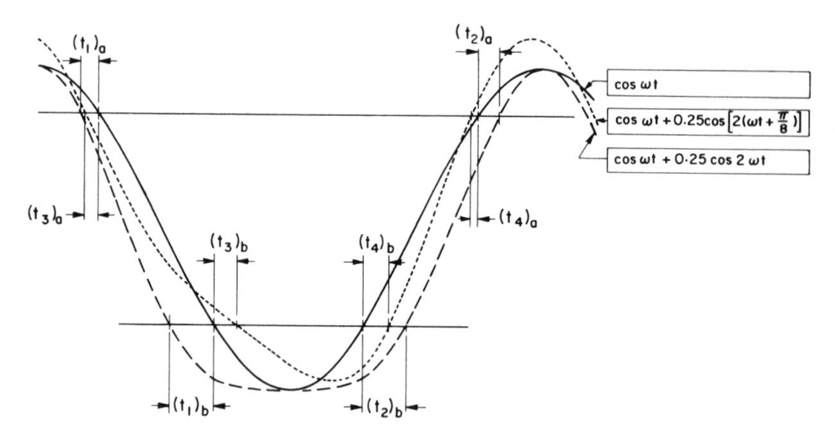

Fig. 8.2. Errors of epoch from waves with harmonics.

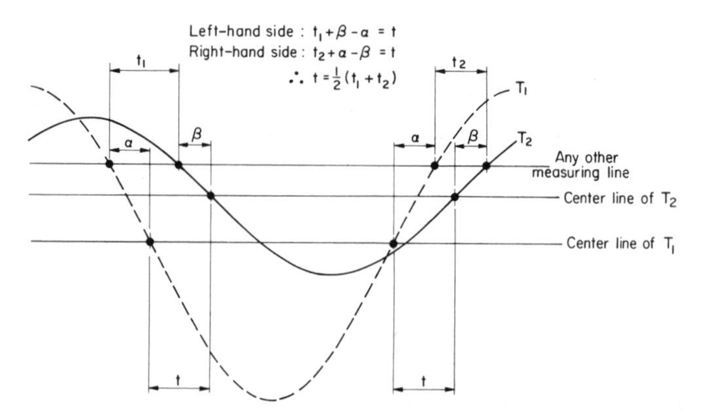

Fig. 8.3. Invariance of phase lag with position of measuring line (pure sine waves).

8.2.1 The Measurement of Temperature or Concentration

Almost invariably the concentration or temperature in a fluid will be examined by a device that produces an electrical signal that is a function of the measured quantity. Concentrations probably will not be detected directly, i.e., by "molecular counting," just as temperatures will not be measured by measuring molecular velocities, but by utilizing some other effect that alters an electrical current or voltage: for example, the thermal conductivity of gases, electrical conductivity of electrolytes, or the light absorption of fluids. Temperatures will alter the electrical resistance of a metal or a solid-state device or it will give rise to a thermoelectric voltage.

Some points to be considered when making a choice are:

(i) The device should be large enough (in dimensions normal to the flow) to average out local space variations. It should be short (e.g., $< \lambda/100$), but long enough to help average out turbulent fluctuations. It should have a response time fast enough not to modify appreciably the concentration signal it is receiving. (A very fast response time will pick up turbulent fluctuations, but damping may be introduced into the electrical signal from the transducer by a shunt capacitance, of size determined by trial.)

(ii) Reproducibility should of course be high, but since nothing is invariable, some check and calibration devices should be incorporated for the highest accuracy.

Good linearity makes subsequent recording and analysis much easier. In general, none of the devices listed above is linear, but for ranges sufficiently small they may be considered so. However, in the case of light absorption, the Lambert–Beer law gives the intensity of monochromatic light emerging from l centimeters of solution of concentration v as being proportional to $\exp(-klv)$, where k is a constant. Thus, it follows from this highly nonlinear relation that the absolute error in v is proportional to the relative error in the measured light intensity, which would be important if a wide range of concentration were used. If, on the other hand, a restricted range of concentrations near $v = 0$ were used, then expansion of the exponential term gives the intensity of the emerging light as approximately equal to $1 - klv$, but there is still a nonlinear relation between relative errors in measured light intensity and concentration.

(iii) Other things being equal, high sensitivity (i.e., a large electrical output for a given change in temperature or concentration) is desirable because noise or electrical pickup in connecting cables and the noise, instability, and lack of linearity in high-gain amplification are less troublesome.

(iv) In some of the devices an electrical emf is produced that can be detected without difficulty by a high-resistance device. In others, e.g., the electrical conductivity cell or resistance thermometer, an electrical current is flowing. This must not be so large that it affects the flowing medium or the measured variable.

In the cases of the electrical conductivity cell and the resistance thermometer the change in current has to be detected in different ways, for a resistance thermometer always has some finite resistance, whereas a conductivity cell has "infinite" resistance at zero concentration. It follows that it is necessary to balance out the resistance of a thermometer at a datum temperature, while for a conductivity cell the current tends to zero at datum (zero) concentration. Hence, the resistance thermometer requires a bridge of resistors as in Figure 8.4(a), while a conductivity cell requires either a similar kind of bridge or a series resistance, the developed emf across which can be applied to a detector as in Figure 8.4(b).

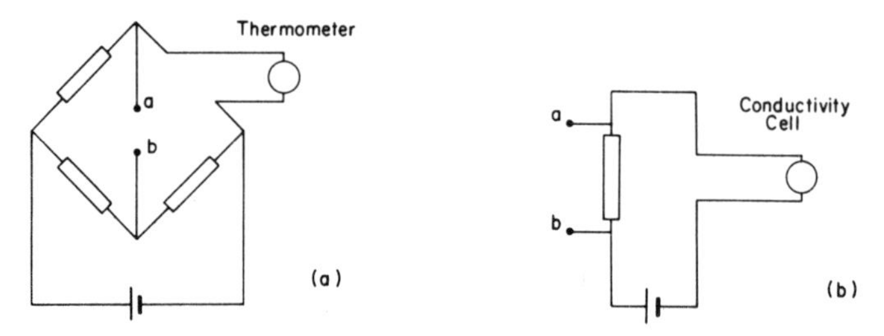

Fig. 8.4. Frequently used circuits of detecting devices; ab is output to detector.

8.2.2 Temperature-Detecting Cells

Thermistor is markedly nonlinear but the response of its electrical network may be made linear at the expense of sensitivity. An electrical handbook will give suitable circuit values.

Electrical Resistance Thermometer Bridge

The analysis of the sensitivity of a Wheatstone bridge has been well treated in the literature. For the special case of a resistance thermometer (with the symbols used in Figure 8.5) the emf E per degree of temperature change is

$$E = \alpha(Wr_1)^{1/2}/\{[(1+n)/n] + [(1+m)r_1/r_g]\}$$

for small changes. A decrease in m and an increase in n beyond unity both have small effects on E (but the current drain and the required battery voltage are increased respectively). Again, the effect of loading the bridge by using finite values of r_g is to reduce E by a factor of $(r_g/r_1)/[1 + (r_g/r_1)]$ times the value of E when $r_g \longrightarrow \infty$. Hence, an equal-armed bridge feeding a high-

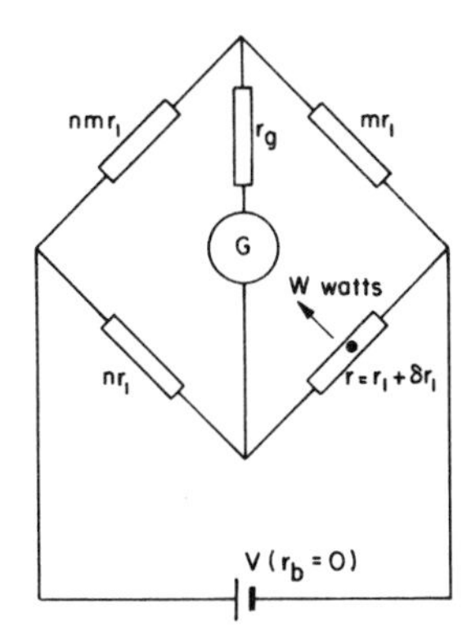

Fig. 8.5. Circuit details of resistance thermometer bridge; r_g = resistance of galvanometer.

resistance detector is desirable. In this case

$$E = \alpha(Wr_1)^{1/2}/2$$

for constant total wattage dissipation, and

$$E = \alpha_1 ir_1/2 = \alpha V/4$$

for constant specific wattage dissipation, i.e., per unit area, where i is the current in the thermometer. The second equation becomes the working formula if the minimum practical wire diameter is specified, the sensitivity of the system being increased by merely using a longer length of wire, increasing V (and so W) in proportion to keep i constant. For any value of R that has been chosen, the ratio of surface area a to the thermal capacity M should be as large as possible in order to endow the thermometer with a fast response. The ratio holds $a/M = 4/dq$, where d is the wire diameter and q is the volumetric specific heat. It is independent of the specific resistance of the wire, although this will, of course, determine the length of wire required.

Leads to the Thermometers: Maximum Resistance

Usually the thermometer requires relatively long connecting leads (as contrasted with the connections to the other resistances in the bridge) and if of copper, these will have a marked temperature sensitivity. The thermometer may be connected in a two-, three-, or four-lead system, as in Figure 8.6(a–c), respectively.

In the arrangement of Figure 8.6(a) any temperature change in the leads will cause a change of output. In the arrangements of Figure 8.6(b, c) only temperature differences between the leads will affect the output of the bridge, and when the resistance of the detector is large, i.e., little current flows, then circuit 8.6(b) is satisfactory.

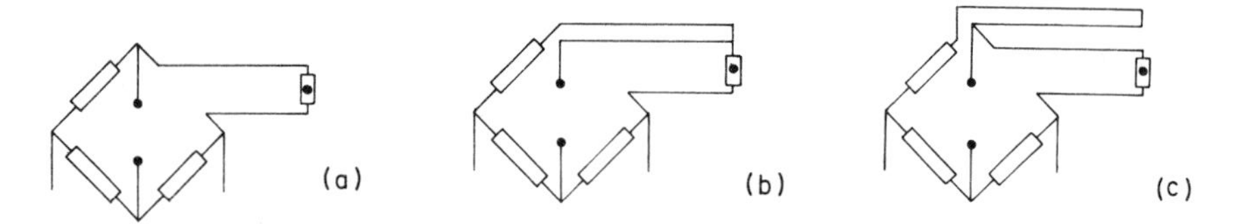

Fig. 8.6. Two-, three-, and four-lead systems.

The methods of choosing the diameter of the leads in Figure 8.6(a, b) can be simply shown; they will demonstrate the superiority of the three-lead system.

Two-Lead System

Let the maximum acceptable uncertainty in the thermometer reading be T_n degrees. This will correspond to an uncertainty of $\Delta R_T = T_n \alpha_T R_T$ ohms in the resistance of any arm of the bridge. The maximum temperature difference between the leads is ΔT_L degrees.

Example 1

Given: $\alpha_L = 4.5 \times 10^{-3}\ \text{deg}^{-1}$, $R_T = 1000$ ohm. Bridge: equal arm. Chosen values: $T_n = 0.001°\text{C}$, $\Delta T_L = 15°\text{C}$.

Hence, $\Delta R_T = 0.001 \times 4.5 \times 10^{-3} \times 1000\ \text{ohm} = 4.5 \times 10^{-3}$ ohm. Now, each set of leads in the bridge could contribute to this. Hence, $\Delta R_L \leq \Delta R_T/4 = 1.125 \times 10^{-3}$ ohm, and so the leads must have a maximum resistance of $R_L = \Delta R_L/(\alpha_L\, \Delta T_L) \leq 1.125 \times 10^{-3}/(15 \times 4.5 \times 10^{-3}) = 1/60$ ohm. The gauge of the connecting leads can then be chosen when their length is known.

Three-Lead System

In this case only the temperature difference between two physically close leads will give rise to the uncertainty, while only half the total length of the leads appears in the calculation of wire gauge.

Example 2

Assume $\Delta T_L = 3°\text{C}$. Then the lead resistance that limits the uncertainty to $0.001°\text{C}$ will be $R_L = \Delta R_L/(\alpha_L\, \Delta T_L) = (1.125 \times 10^{-3})/(3 \times 4.5 \times 10^{-3}) = 1/12$ ohm.

8.3 PRACTICAL ASPECTS OF THE GENERATION OF SINE WAVES

8.3.1 General

It is difficult or impossible to generate really pure, steady sine waves, but in these systems the higher harmonics are attenuated more rapidly than the fundamental. Unfortunately, the second harmonic is the one usually present in large amounts and this is often attenuated so slowly (particularly at low fundamental frequencies) that the fundamental is largely lost as well.

8.3.2 Generation of Power Sine Waves

The final source of heat has usually been an electric heater in the flowing stream. In Figure 8.7(a) this is shown as R, being supplied from a power source P, which in turn is controlled by the program generator G to produce a cyclic wave. Usually this wave is a sinusoidal voltage (but as will be seen, it is possible to produce a sinusoidal wave of power). So the current for constant R is

$$i(t) = I[(\cos \omega t) + b], \qquad b = \text{any value} \tag{8.2}$$

where I is the semiamplitude and Ib is the standing value at the midpoint of the swing of v (the programming voltage).

The power output $W(t) = i^2R$ is

$$W(t) = \tfrac{1}{2}RI^2[4b(\cos \omega t) + (\cos 2\omega t) + 1 + 2b^2] \tag{8.3}$$

Thus, the power—and hence temperature—wave will contain first and second harmonics, along with a standing value. The relative magnitude of all these quantities will depend on the ratio b.

In practice the relation between the output current I of the power unit and the programming voltage v of the program generator may be linear or nonlinear, as shown in Figure 8.7(b). If it is linear, then the only parameter that can be varied is b, but this has a profound effect on the composition of the wave, as can be seen from Equation (8.3). With the use of this equation, Figure 8.8 demonstrates the shape of the power (temperature) wave as b is varied. It is assumed that the relation between output current i and control voltage v is linear (line I), and that the corresponding square-law relation between power and control voltage would be given by line II; the control signal may be biased so that the midpoint of the swing can take up various positions relative to these curves. Figure 8.8(i)–(iv) shows examples and the

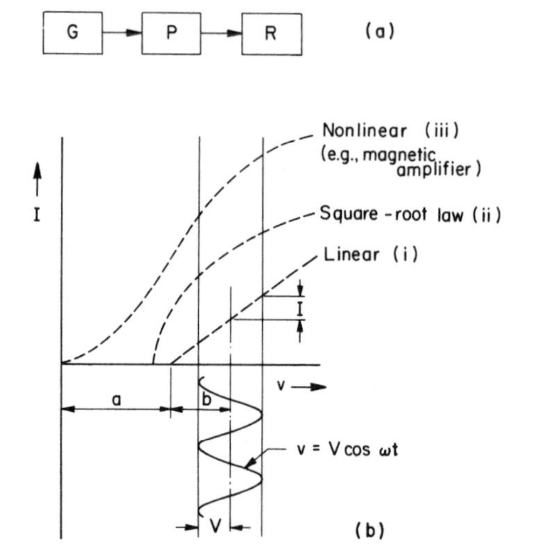

Fig. 8.7. Temperature sine-wave generator; typical behavior.

corresponding waves as deduced from Equation (8.3). Position (i) produces a pure power wave of new frequency 2ω, for now $b = 0$, and so Equation (8.2) gives

$$W(t) = \tfrac{1}{2}RI^2(\cos 2\omega t + 1) \tag{8.4}$$

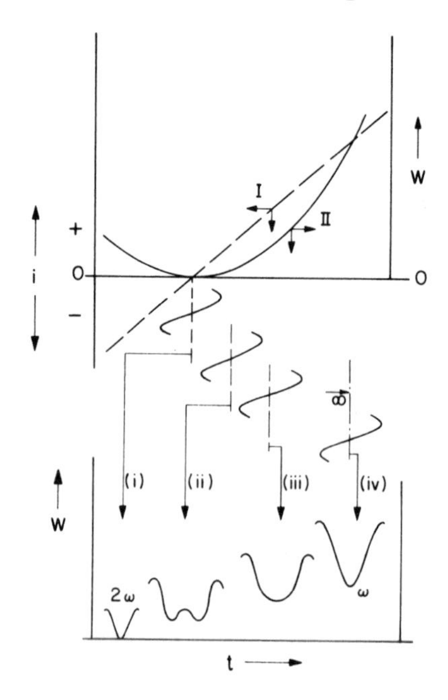

Fig. 8.8. The dependence of the composition of the temperature wave on the biasing of the control signal.

The device thus acts as a *frequency doubler*, which is potentially a useful attribute. The amplifier and controller must both have stable characteristics, for a slight drift to position (ii) results in subharmonic (frequency ω) as shown. This latter situation is undesirable, for, as discussed earlier, harmonics or subharmonics can lead to incorrect measurements of phase lag and amplitude. In particular, the subharmonics would be attenuated *less* than the fundamental as the wave passes through the system, and so the wave would become less pure as it progresses.

In position (iii) the value of b is larger and the output, described by Equation (8.3), may be thought of as being of frequency ω but with a large amount of second harmonic.

In the limiting position, indicated by (iv), $b \longrightarrow \infty$ and the equation becomes

$$W(t) \approx \tfrac{1}{2}RI^2[4b(\cos \omega t) + 2b^2] \tag{8.5}$$

That is, it tends to a pure wave but with a very large amplitude and a large standing value. Both these attributes would necessitate a large power unit, with perhaps high temperatures.

Example 3

A power unit used as a frequency-doubler drifts so that $b = 0.01$, i.e., the midpoint is displaced by $I/100$. Deduce the effect on the swing of the resultant wave.

Answer

From Equation (8.3)

$$W(t) = \tfrac{1}{2}RI^2[0.04(\cos \omega t) + (\cos 2\omega t) + 1 + 2 \times 10^{-4}]$$

All minima [as illustrated in Figure 8.8(ii)] lie on the line $W(t) = 0$, while each maximum is alternately raised and lowered by 0.04 in a total swing of 2, i.e., by 2%.

Linearization of the Power Output

If the current output is proportional to the square root of the control voltage, as line (ii) in Figure 8.7(b), then the power output into a constant load will be linear, and drift of the control voltage would not result in an impure wave.

Line (iii) shows a nonlinear response that certain types of amplifiers might have. Some magnetic amplifiers behave like this and the shape of the curve can be altered by suitable feedback. By arranging that the control voltage (or current, in the case of a magnetic amplifier) covers a suitable part of the curve, a reasonably linear power output is obtainable.

For the case where the current output of the amplifier is linear the power output may be linearized by arranging negative feedback proportional to the power output as in Figure 8.9. With the symbols there the effect may be computed, for

$$V = G_E v_e, \qquad v_e = v_a - v_m, \qquad v_m = G_M W$$

and so

$$v_a = (V/G_E) + G_M W$$

Hence, $v_a \approx G_M W$ if $G_M W \gg V/G_E$, i.e., if $G_M G_E \gg V/W = 1/I\ A^{-1}$.

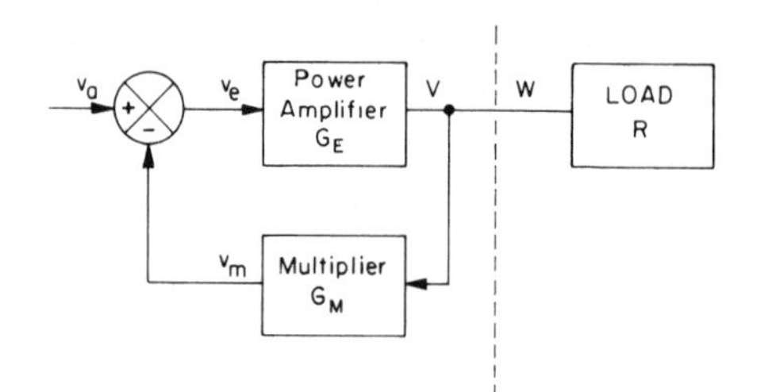

Fig. 8.9. Block diagram of a multiplier to linearize the output of a power unit.

Example 4

Using the following values (taken from a commercial unit), compute the amplification around the loop to provide reasonable linearity over nine-tenths of the power unit's output: $G_M = 0.2$ V/1400 W, $G_E = 36$ V output for 2 V control voltage; hence, $G_E = 36/2$. So, $G_M G_E = 2.5 \times 10^{-3}\ A^{-1}$.

Answer

Now, for there to be linearity over, say, nine-tenths of the total range of the power unit (viz., 0–10 A) it must be specified that $G_M G_E \gg 1/I_{min}$, where I_{min} is the lower end of the output range over which linearity is desired. Thus, in the present case $I_{min} = (1/10)(10) = 1$ and so $G_M G_E$ must be greater than 1/1. But as stated, $G_M G_E = 2.5 \times 10^{-3}\ A^{-1}$. Hence, an additional amplification stage, having a gain of much greater than $1/(2.5 \times 10^{-3}) = 400$, is required in order to satisfy the condition for linearity over nine-tenths of the output; this latter gain is to be dimensionless, i.e., a voltage amplifier is required.

This gain in linearity is at the expense of requiring a greater value of v_a to produce the full output of the power unit, and the greater the value of $G_E G_M$, the greater the requirements on v_a, as shown in Figure 8.10, which illustrates how the output power in watts depends on the variable $G_E v_a / R^{1/2}$; this dependence is governed by the parameter $A = G_M G_E / R^{1/2}$; as it increases, the approximately linear portion of the curve stretches out to span a greater range of power, but requires a greater value of the variable $G_E v_a / R^{1/2}$ to

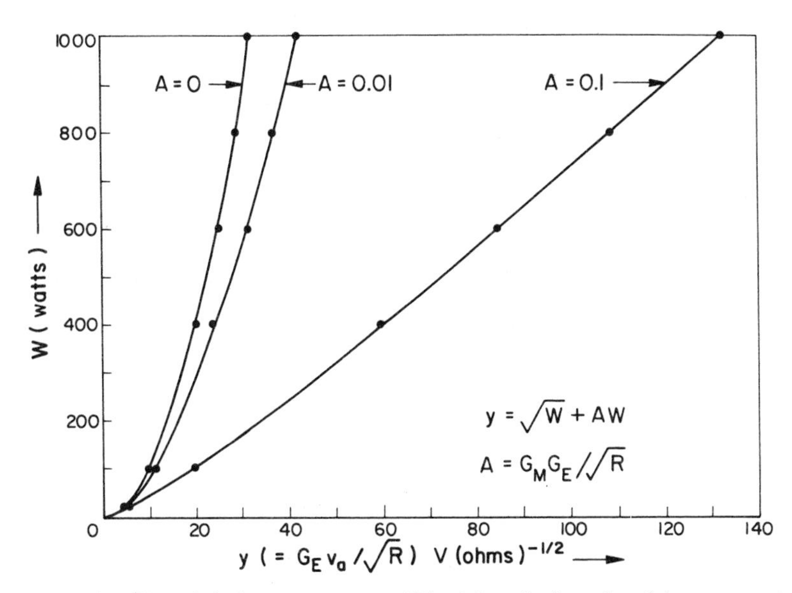

Fig. 8.10. The effect of the loop parameters of Fig. 8.9 on the linearity of the power output.

achieve this. It is assumed that the value of R is fixed by the choice of power unit, so the last-named variable may be increased by increasing G_E; this would save the necessity for increasing v_a. Practically, it means connecting a voltage amplifier before the power amplifier. Turner and Goss [1] report the use of a Hall-effect multiplier.

General

The repeated mention of the effect of drift stresses the desirability of a stable control-voltage source and power amplifier having a linearized power output if pure, steady sine waves are to be generated. Additionally, a steady, controlled temperature upstream of the sine-wave heater is needed, the sine wave being superimposed. This controller would have a feedback loop, and it may have derivative action. These two attributes could also be introduced into the sine-wave generator, but an attempt to use a proportional control feedback (utilizing another sine-wave voltage source of variable phase lag as desired value) was reported by Goss [2] to be not very successful.

Derivative action with a resistance thermometer could be obtained by a differentiating network in the bridge output, but an alternative, giving both derivative and proportional signals, could be as in Figure 8.11; thermometer *a* gives the proportional signal, thermometers *b* and *c* the derivative. If the distance Δl between thermometers *b* and *c* is small, then the portion of the signal E' from the bridge due to a time rate of change of temperature of

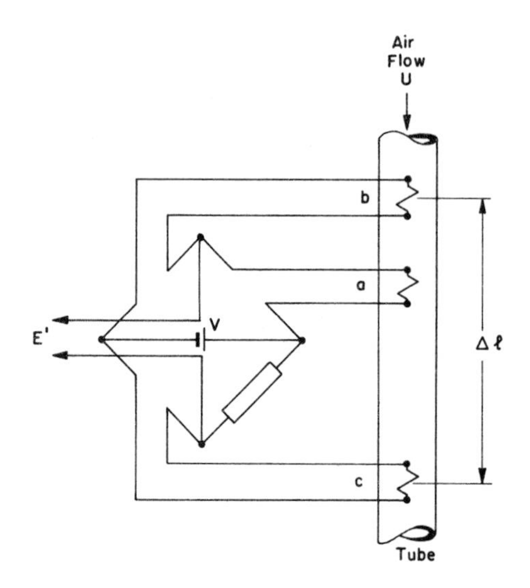

Fig. 8.11. Three thermometers used to provide proportional plus derivative output.

dT/dt is

$$E' \approx (\alpha V \, \Delta l/4U)(dT/dt)$$

where V is the bridge voltage and U is the average velocity of the fluid past the thermometers.

Example 5

$$\alpha = 4 \times 10^{-3} \text{ deg}^{-1}; \; V = 25 \text{ V}; \; U = 1000 \text{ cm sec}^{-1}; \; \Delta l = 5 \text{ cm}$$

Then, for $dT/dt = 1 \text{ deg sec}^{-1}$, $E' = [(4 \times 10^{-3} \times 25 \times 5)/(4 \times 1000)] \times 1/1$, i.e., $E' = 0.25$ mV.

If used in the sine-wave generator, then let the sinusoidal temperature wave produce a signal $A_b \sin \omega t$ from b alone, and $A_c \sin[\omega(t - t_{lag})]$ from c alone, where t_{lag} is the phase lag (distance–velocity lag) between the two. The resultant signal at any time is

$$y = \pm\{A_b \sin \omega t - A_c \sin[\omega(t - t_{lag})]\} \tag{8.6}$$

where the $\pm$ sign is put in here only to indicate that the bridge may be balanced to give either a negative or a positive signal; i.e., either one of two signals out of phase by π radians.

The various cases where A_b and A_c are the same or different, or when t_{lag} is zero or finite, can be examined by trigonometric expansion of Equation (8.6), but the important cases are as follows.

(i) $A_b \neq A_c$; $t_{lag} \neq 0$; that is, the thermometers are not identical. Then,

$$y = Y(\sin \omega t \cos \psi + \cos \omega t \sin \psi)$$

where $\tan \psi = (\sin \omega t_{lag})/[(A_b/A_c) - \cos \omega t_{lag}]$ and

$$Y^2 = A_b^2\{[1 - (A_c/A_b)]^2 + 2(A_c/A_b)(1 - \cos \omega t_{lag})\}$$

(ii) $A_b = A_c = A$ and $t_{lag} \neq 0$; that is, the thermometers have identical responses. Then,

$$y = 2A \sin (\omega t_{lag}/2) \cos [\omega t - (\omega t_{lag}/2)]$$

If ωt_{lag} is small, e.g., $\omega t_{lag} < 0.2$ rad, and $t_{lag} = \Delta l/U$, then

$$y \approx A\omega(\Delta l/U) \cos [\omega t - (\omega t_{lag}/2)]$$

and for this special case the signal is approximately $\pi/2$ radians in advance of the temperature wave at thermometer a; i.e., it is a differentiating device with the amplitude of the signal being proportional to the frequency. Thus, if the wave is impure, higher harmonics would give a larger proportionate signal.

No Separate Programmer

Some workers have controlled the power by a linear rheostat (a variable resistor) whose sliding contact is driven with simple harmonic motion. The circuit is as in Figure 8.12(a). The current $I = V/(R + r)$ [with the symbols

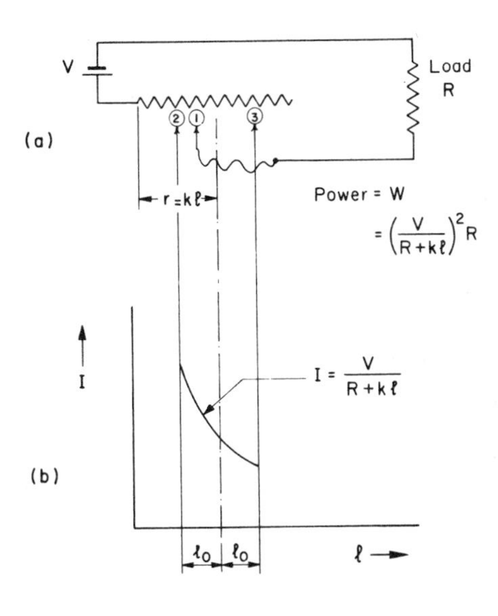

Fig. 8.12. (a) Circuit to generate a periodic current from a rheostat. (Contact 1 oscillates between positions 2 and 3.) (b) The current as a function of displacement of the rheostat contact.

as shown in Figure 8.12(a)] would vary with l as shown in Figure 8.12(b). This curve is convex to the l (i.e., time) axis, and so, the power being proportional to I^2, would produce a very impure wave. If a circuit such as Figure 8.13 is used, where r is a linear potentiometer, then unless $R \gg r$, the output

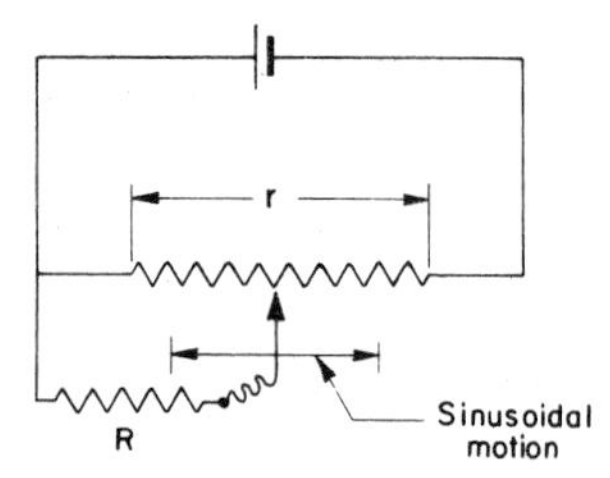

Fig. 8.13. Potentiometer used to generate periodic waves.

is nonlinear (see the section below on potentiometer loading) and if R is made large relative to r, then most of the heat will be dissipated in the potentiometer. There is usually some difficulty brought about by the heavy gauge of wire used in r; the output is in discrete steps and at peak and trough an appreciable "flat spot" is produced for this reason.

Variable Transformer

If in Figure 8.13, r denotes a variable transformer instead of a potentiometer, there is much less or no difficulty about the loading effect of the resistance R, but since r is a power device, the winding must be of heavy gauge and the difficulties mentioned in the last paragraph remain.

8.3.3 The Initial Generation of Sine Waves—The Programmer

The recommended way of generating precise *temperature* sine waves in a flow system is to separate the programming and the power producing functions; the programmer then has to produce, say, a sinusoidally varying voltage but little power.

The generation of waves of *concentration* may be more difficult. Often a mechanical oscillation which controls concentration by altering flow rates is used. A few devices will be described below.

In temperature waves the initial wave of voltage may be obtained from a purely electrical source or from a mechanical device. These, of course, have to be driven at a speed that is constant over both the short term and the long term. These devices are either reciprocating or rotary.

1. The reciprocating motion is usually from a Scotch yoke arrangement, as in Figure 8.14, which produces it from rotary motion. Mechanical imperfections in the bearings (a), the pin (b), and the slot faces (c) will make the velocity variations not the pure sinusoid that should be produced, especially at peak and trough. Continuous unidirectional loading so that only one of the surfaces (c) is used and the use of "linear" (ball) bearings will minimize the irregularity at peak and trough. The reciprocating motion is then used to drive some other device that actually gives heat or matter.

2. Rotary motion is mechanically simpler. It may cause a sinusoidal voltage in several ways.

(i) Figure 8.15 shows electrodes revolving in an electrolytic cell. The electrodes *c* and *d* should be flat, parallel, and large so as to give a uniform

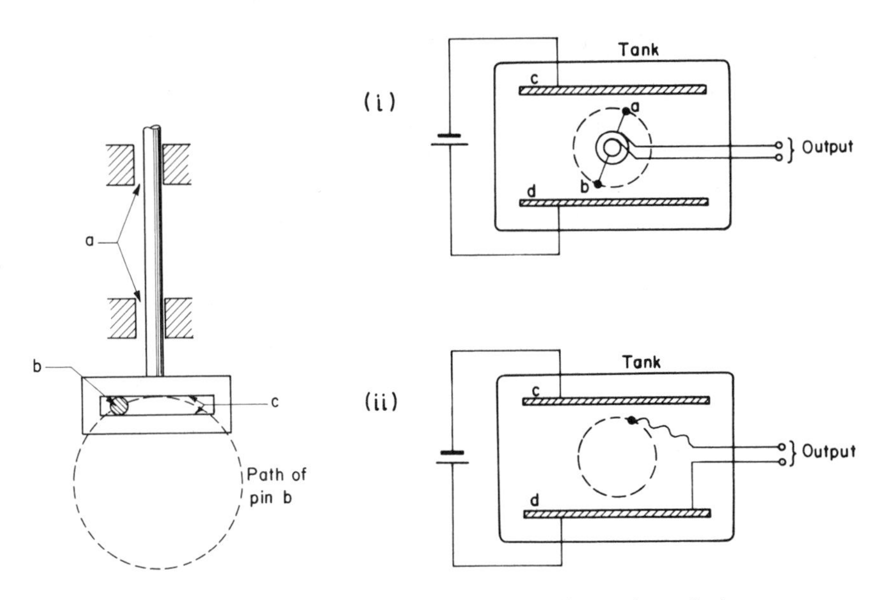

Fig. 8.14. Scotch yoke and tolerances.

Fig. 8.15. Generation of sinusoidal emf by electrolytic cells. (i) Two revolving electrodes. (ii) One revolving electrode.

potential field. The component of motion normal to the equipotential lines produces the sinusoidal variation in potential. If a current is taken from this potentiometer, it must be remembered that unless two revolving electrodes are used, the internal resistance between one revolving electrode and the flat electrode will depend upon the position of the electrode. The electrodes must be of small diameter and of a material such that there is no overvoltage or polarization between them and the electrolyte. The current must be small (preferably microamps per square centimeter), for otherwise, depletion of the electrolyte leads to nonuniformity of the equipotential surfaces. In any case the electrolyte should be stirred continuously and kept at a closely controlled temperature—the conductivity of electrolytes is highly dependent on temperature. Thus, a device that is simple in principle can become quite elaborate if it is to be at all precise. Rawcliffe [3] has investigated these devices.

(ii) Sine–cosine potentiometer. This usually refers to a wire-wound potentiometer. Two types are used. The commoner has a wire-wound card shaped so the resistance is proportional to the sine or cosine of a displacement. For continuous rotation the card is bent into the form of a circle and the resistance is that between the instantaneous position of the revolving contact and a fixed tapping, as in Figure 8.16. The other type has a flat, rectangular card wound with resistance wire. An electrical contact describes a circular track on a face of the wound card, as in Figure 8.17. It is thus similar to the electrolytic cell of (i). In either case the greater the number of turns of wire encompassed in a revolution, the greater the *resolution*; that is, the smaller is each step in the resistance, but if fine wire is used, the shorter will be the life of the winding. The resistance has to be chosen to suit the circuit.

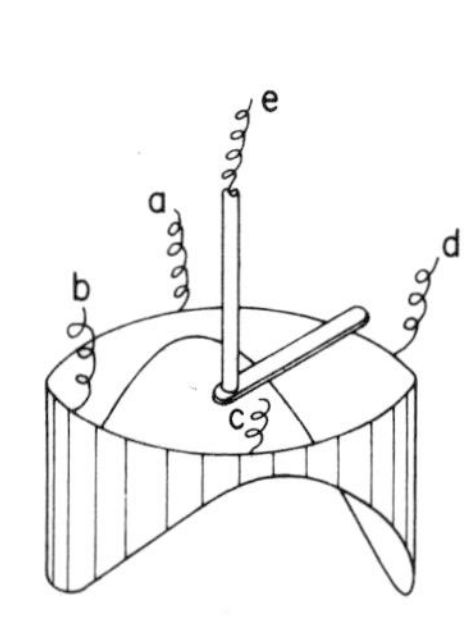

Fig. 8.16. Generator of sinusoidal emf; shaped card potentiometer. (Connections are labeled to correspond to Figure 8.18. Shape illustrative only.)

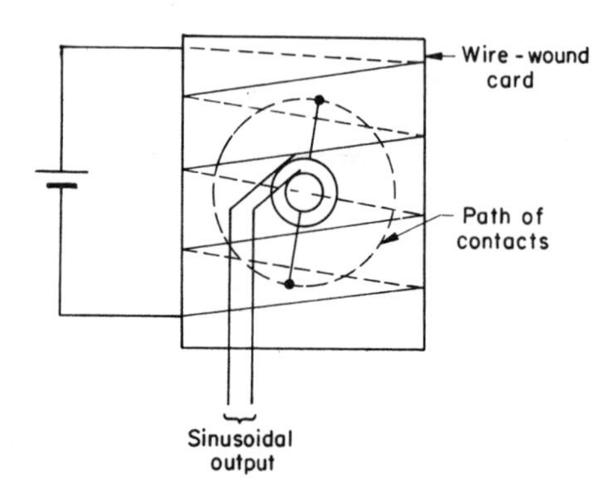

Fig. 8.17. Generator of sinusoidal emf; rectangular card potentiometer.

Loading of Potentiometers

The sinusoidally varying resistance of a sine–cosine potentiometer will only produce a potential that is also a pure sine wave (i.e., one that has no harmonics introduced into it) if the output is fed into a device (the *load*) that has infinite electrical resistance. In practice this resistance is finite, although it may in fact be of the order of megohms; there is a complementary problem in that very often the load requires the source (in this case the sine–cosine potentiometer) not to exceed a certain value. It is possible to satisfy these two requirements with the circuit shown in Figure 8.18, which satisfies simultaneously the conditions that (i) the load resistance across the potentio-

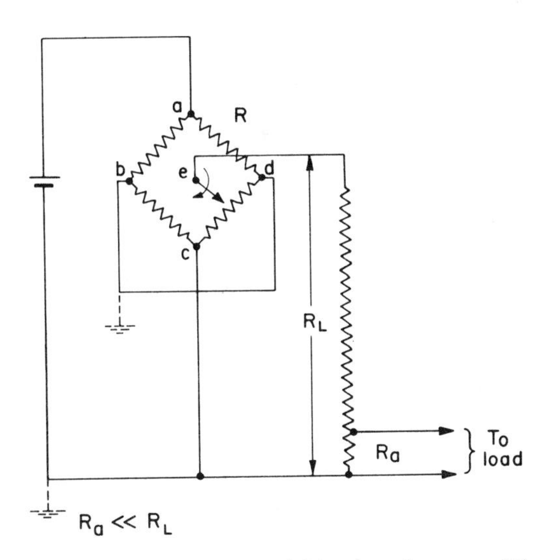

Fig. 8.18. Potentiometer circuit satisfying impedance conditions.

meter is high, (ii) the source resistance feeding the load is low, and (iii) the resistance of the potentiometer may be high enough to give good resolution.

Effect of Loading on the Output of a Potentiometer

The calculation of the departure from linearity of the output voltage caused by a finite load resistance is described in the literature—for example, Kaufmann [4]—while the error curve for the circuit shown in Figure 8.19 is plotted as a function of the resistance r and is of the general form shown in Figure 8.20. It may be used to determine graphically the shape of the output voltage wave across R_L by the construction shown in Figure 8.21.

The left-hand portion in this figure shows a plot of the relative error e/R as a function of the relative resistance of the potentiometer r/R. The coordinates are at 45°. The right-hand portion of the diagram contains two plots: (i) r/R against an angle which may be thought of as the angle of rotation of the shaft of a circular potentiometer, or as the quantity ωt; and (ii) the actual voltage across the load R_L (as a fraction of the voltage across the potentiometer, viz., V_L/V_R).

Thus, for a sine–cosine potentiometer the full line on the right-hand side represents the relative resistance of the potentiometer as the shaft revolves, while the dashed line indicates the voltage that results across load R_L, and which perhaps then programs a power unit. To obtain this latter curve, the construction is (as shown) to draw a line from point a at any deflection to arrive at point b at the same deflection.

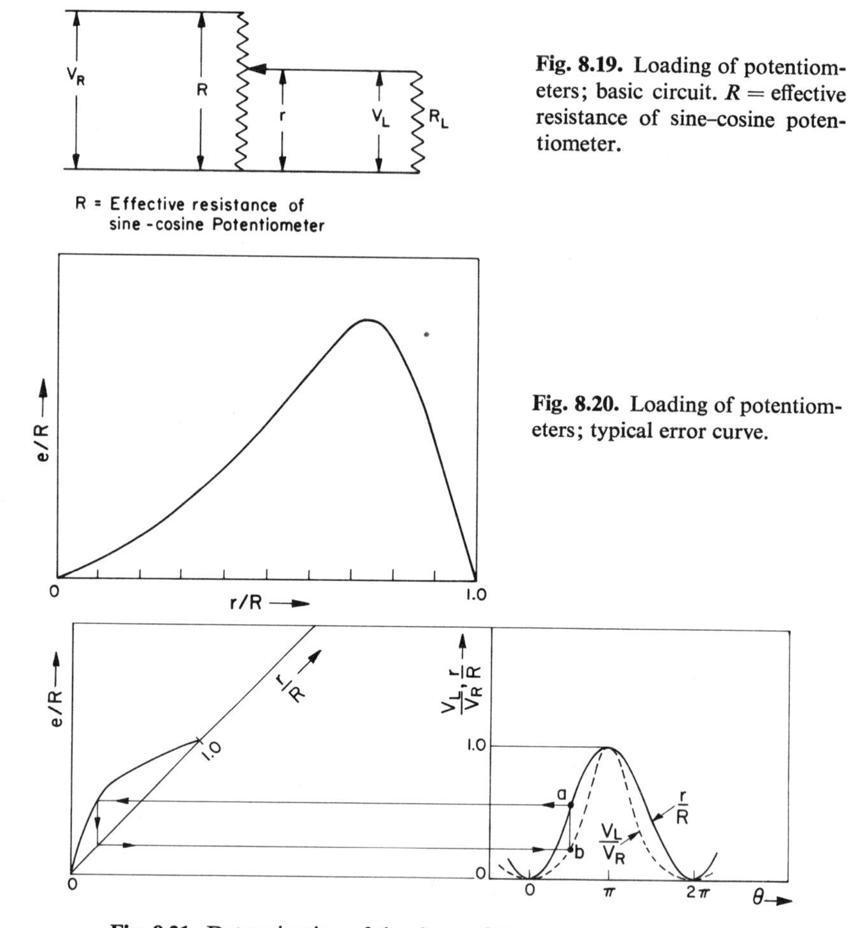

Fig. 8.19. Loading of potentiometers; basic circuit. $R =$ effective resistance of sine–cosine potentiometer.

Fig. 8.20. Loading of potentiometers; typical error curve.

Fig. 8.21. Determination of the shape of the output voltage wave.

Alternatively, the value of the output voltage may be calculated, using the equation in Kaufmann [4]. The amounts of the harmonics present could then be calculated if desired.

8.3.4 The Final Output Device

This should not introduce its own distortions; i.e., it should have a fast response and should produce a uniform concentration normal to the flow and be linear. For *mass concentration waves*, in general (although it could be used for temperature waves), the final device produces a mixing of two streams of different compositions as in Figure 8.22. For either stream, if the volume of a reservoir is

$$\tilde{v} = \tilde{V}_0 + \tilde{V} \sin \omega t$$

the downstream rate of flow is

$$q = Q_{\text{in}} - (d\tilde{v}/dt) = Q_{\text{in}} - \omega\tilde{V} \cos \omega t$$

Hence, the total rate of flow into the mixing region M is

$$\begin{aligned} q_1 + q_2 &= (Q_{\text{in}})_1 + (Q_{\text{in}})_2 - \omega(\tilde{V}_1 + \tilde{V}_2) \cos \omega t \\ &= \text{const} = Q, \quad \text{if} \quad \tilde{V}_1 = -\tilde{V}_2 \end{aligned}$$

i.e., if the amplitudes of the reservoir changes are the same and π radians out of phase. The amplitude of the concentration into the mixer M is (if the concentration in, say, stream 2 is taken as datum, if the concentration in stream 1 is constant at v_1, and if the above condition holds)

$$v(t) = [Q_1 - \omega\tilde{V}_1 \cos(\omega t)]v_1/Q$$

which is a sinusoid with a mean of $Q_1 v_1/Q$ and an amplitude of $\omega\tilde{V}_1 v_1/Q$, having a maximum of $v_1/2$ which occurs when $\omega\tilde{V}_1/Q = \frac{1}{2}$.

Example 6

Find the amplitude of the reservoir volume in each stream to give the maximum theoretical amplitude when the flow rate is 2 liters sec^{-1} and the radial frequencies are: (i) $\omega = 0.1$, (ii) $\omega = 1.0$, (iii) $\omega = 10.0$ rad sec^{-1}.

Answer

The values of $\tilde{V}$ are given by $\tilde{V}_1 = \tilde{V}_2 = \frac{1}{2}Q/\omega$. Thus, the values are: (i) $\tilde{V} = \frac{1}{2} \times 2/0.1 = 10$ liters, (ii) $\tilde{V} = 1$ liter, (iii) $\tilde{V} = 0.1$ liter.

These results are interesting because, unlike most attributes of time-varying measurements systems, a higher frequency has a perhaps less stringent requirement, viz., a smaller volume swing.

An alternative final output device is one to control the flow; thus, for example, Deisler and Wilhelm [5] operated valves (i.e., variable restrictions) in the supply lines as in Figure 8.23(a), while Turner [6] used a variable head of liquid as shown in Figure 8.23(b). This shows two siphons alternately raised and lowered relative to liquids of two different concentrations. Flow in the capillary tube siphon was laminar, hence the flow was linearly proportional to the head.

The Mixer

The above methods of generating sinusoids, by bringing together two streams, require a mixing process that is represented by M in Figures 8.22, 8.23(a), and 8.23(b). This must (a) provide good mixing to prevent unwanted

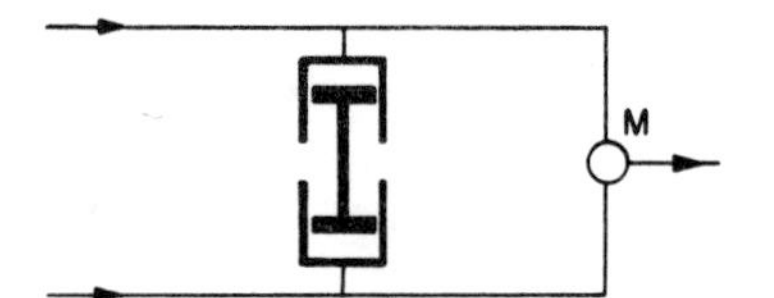

Fig. 8.22. The final output device using reservoirs of varying capacity.

variations, (b) be of small size to reduce the attenuation that is always caused by a mixer. If, for the purpose of design, M is assumed to be a "perfect mixer" (which is defined as one in which the output concentration is always the same as that in any part of the mixer), then Kramers and Alberda [7] give the ratio of the amplitudes as

$$V_{out}/V_{in} = 1/[1 + (\omega\tilde{V}_M/Q)^2]^{1/2}$$

where $\tilde{V}_M$ is the volume of mixer.

Example 7

Find the ratio of amplitudes of sine waves before and after a perfect mixer for the two extreme frequencies given in Example 6 when the volume of the mixer is: (a) $\tilde{V}_M = 0.1$, (*b*) $\tilde{V}_M = 1.0$ liter.

Answer

CASE (i) $\omega = 0.1$

(i) (a) $\tilde{V}_M = 0.1$. Thus, $\omega\tilde{V}_M/Q = 0.1 \times 0.1/2 = 0.5 \times 10^{-2}$ (dimensionless). Hence, $V_{out}/V_{in} \approx 1$.

(i) (b) $\tilde{V}_M = 1.0$. Thus, $\omega\tilde{V}_M/Q = 0.5 \times 10^{-1} = 5 \times 10^{-2}$. Hence, $V_{out}/V_{in} \approx 1$.

CASE (ii) $\omega = 10.0$

(ii) (a) $\tilde{V}_M = 0.1$. Thus, $\omega\tilde{V}_M/Q = 10 \times 0.1/2 = 0.5$. Hence, $V_{out}/V_{in} = 1/(1.25)^{1/2} = 0.9$.

(ii) (b) $\tilde{V}_M = 1.0$. Thus, $\omega\tilde{V}_M/Q = 5$. Hence, $V_{out}/V_{in} = 1/(26)^{1/2} = 0.196$.

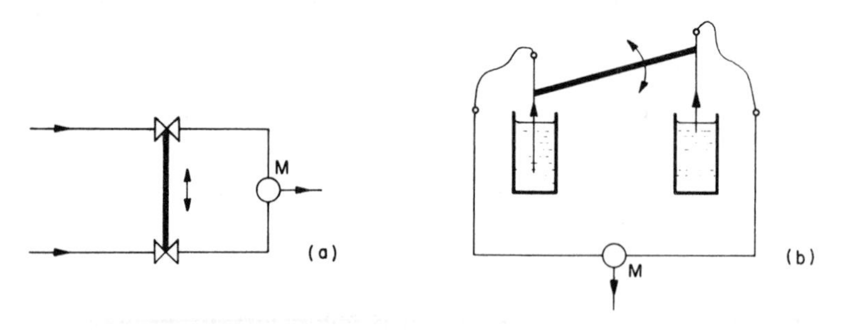

Fig. 8.23. Final output devices that operate by controlling the flow.

This example serves to stress the importance of having only a small mixing volume at higher frequencies.

For *temperature waves* the heater will generally be an electrical heater; it should be made of material whose resistance does not vary appreciably with temperature—which would introduce harmonics. Thus, one of the common resistance wires may be chosen, having regard to the ease of making connections (i.e., its solderability). The speed of response is a function of the surface area of the wire, its thickness, and of the thermal inertia of any supports that may be necessary. The thinner the wire, the faster the response, but the higher the voltage across the heater for a given output, $W = E^2/R$. This may be undesirably high or may not be available from a chosen power unit.

There are conflicting constraints—the correct resistance must be married to sufficient surface area, to prevent the heater temperature becoming too high, while the smaller the mass, the less the attenuation of the energy wave. The connecting of heaters in parallel can satisfy these conditions. Table 8.1 shows how the total mass decreases for a fixed surface area and a fixed resistance.

TABLE 8.1

MASS OF m ELECTRICAL HEATERS IN PARALLEL[a]

m	Wire diameter	Mass of heater	Total length of wire
1	1	1	1
2	1/1.59	0.63	1.59
3	1/2.08	0.48	2.08
4	1/2.52	0.40	2.52
8	1/4.00	0.25	4.00
12	1/5.24	0.19	5.24
16	1/6.35	0.16	6.35

[a]To give the same surface area. Values given are relative to the case where $m = 1$.

Example 8

A total of 100 W is to be dissipated in a heater. The maximum voltage is to be 50 V and the *chosen* surface area of wire is 1500 cm^2 (giving 0.066 W cm^{-2}). The resistance required is $(50)^2/100 = 25$ ohms. For a single unit a wire gauge of 20 B&S is found by trial to have both the required surface area and resistance. Then, 58.8 cm of wire are needed and the weight of bare wire is 259 g.

On the other hand, Table 8.1 shows that if, say, eight sections are used, then the same area and resistance result from using wire of one-quarter this thickness (i.e., 32 B&S gauge) and one quarter the mass, viz., 65 g, with a corresponding improvement in response.

8.3.5 The Stability and Precision of the Frequency

Most of the above methods of generating a sine wave involve mechanical rotation. Usually a synchronous electric motor, using the electrical supply mains, drives gears. When the highest precision is desired the possible errors introduced may be significant. The frequency of the electrical supply may be different from what is assumed and it may vary with time. Again, sloppiness in the bearings or errors in the gears can introduce irregularities. Each tooth in a gear wheel meshing with another one pushes a tooth of the latter through a finite distance. Unless each tooth is of perfect involute shape, with no clearance, each driven tooth is moved at a speed that varies slightly over part of its travel. That is, there is a "ripple" in the movement which is translated into a higher harmonic. For this reason, the highest wave purity requires gears of high precision. The larger the diametral pitch (the number of teeth per inch of diameter), the higher, and hence less objectionable, is the frequency of ripple.

The situation is probably worse when the popular worm reduction gear is used. The long helical tooth on the worm (driving a pinion) has inaccuracies that produce irregularities in the speed of the output shaft. The type in which the worm is "coned" (i.e., the diameter varies along the length, to mate with the curvature of an arc of the pinion) is likely to be better, for a number of teeth of the pinion are simultaneously meshing with those on the worm. Furthermore, the worm may have more than one "tooth"—called a "start."

Motor Speed

It may be necessary to drive a synchronous motor from a power amplifier that in turn is driven by a stable oscillator.

The stability of a direct current motor with feedback speed control is probably not high enough.

Measurement of Speed or Frequency

Both measurements are the same in principle—the finding of a time between two events. This event may be (for a sine wave) a specified height of a recorder trace, or a value of an electrical signal that is proportional to a temperature, or (for a rotational speed) the position of a cam (that perhaps actuates a micro switch) or the intensity of light reflected from a mark on a shaft blocked by a tooth or passing through a hole in a revolving wheel. If the time recorded is at an instant when the values of this quantity are inconstant from event to event, then the time interval will, of course, be in doubt. The problem is then the same as that involved in sampling in which a time-changing quantity has to be measured. In Figure 8.24, if the device that

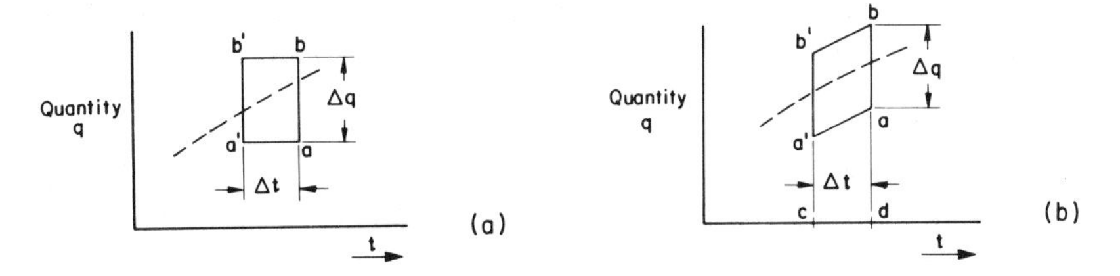

Fig. 8.24. Possible uncertainty in sampling a time-varying quantity.

measures or records is uncertain to within a certain range *ab*, having a mean value that is either fixed as in Figure 8.24(a) or is dependent on the time as in Figure 8.24(b), and if the time at which the actual sampling is done is undertain over the time interval *aa′*, then the sample value may be anything within the boxed areas.

In the above, "uncertainty" would have to be defined in statistical terms for any further analysis or quantitative development. For example, the probability surface could be drawn within the area *aba′b′*, being zero at the boundary.

Reduced-Frequency Sampling

The experimental problems associated with the uncertainty in the sampling process and with the frequency response of the whole measuring device tend, in practice, to merge. This is especially true when a portion of the flowing medium is diverted or removed from the mainstream, for, apart from the behavior of the sensor itself, there tends to be mixing in the conveying line.

Keyes [8], faced with this problem, designed a rapid sampling valve which quickly removed a small sample of gas and allowed it to flow to a fast-response concentration (thermal conductivity) cell. The interesting feature was that the sampling was done at intervals of $n\tilde{t} + (\tilde{t}/m)$ as shown in Figure 8.25. The integers n and m may be independently varied. Each sample differs in concentration by only a small amount; i.e., effectively, the flow along the sampling tube has only a low-frequency variation. A recorder, for example,

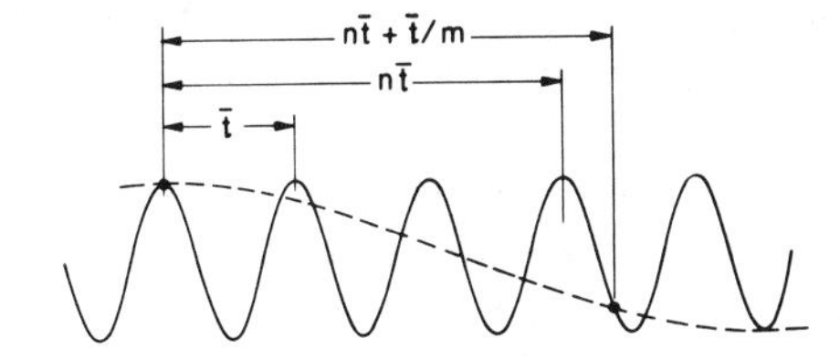

Fig. 8.25. Method used by Keyes [8] for reduced frequency sampling.

thus shows the amplitudes and phase angle of the input and output waves, but at a much reduced frequency. The principle is similar to that of the stroboscope, used to "stop" or "slow" mechanical motion.

8.4 THE FLOW OF THE FLUID

The convective effect of the fluid is central to the processes that are described in this book (although the actual value is often not required). That is to say, the flow must be as the mathematical model describes it. Usually, the model assumes the flow to be constant in time and, especially for a packed bed, uniform in radial position. Short-range or short-time variations must be small compared with reference dimensions, e.g., the size of a bed or measuring device, and of the period of the wave, for example. Achieving these states with a high degree of precision is usually difficult.

Producing a Constant Flow

Commercial controllers will not usually give the degree of precision that is required. However in a given gas flow system, if the temperature of the gas is kept constant (apart from the transient or cyclic temperature), then if the gas pressure is kept constant also, the flow rate will remain constant to within the same tolerances as are the temperature and pressure. For gases a bubbler (which is a pipe of ample diameter dipping into a liquid in a vessel and connected to the gas supply line just before or just after the temperature controller) can keep the pressure constant to within 0.1 cm water gauge. (It is adjusted so that a small stream of bubbles is always flowing; any increase or decrease in pressure will cause an increase or decrease in the gas bypassing the flow tube, and the pressure will be kept steady to within the above limits.) For liquids, a vertical tube connected to the supply line, again near the temperature controller, will keep the pressure constant if there is always a small overflow.

Getting fluid flow uniform radially may also be difficult—there is always a great tendency for any nonuniform velocity profile or rotation to persist even through packed beds. Vanes, eggboxes, honeycombs, and turbulence controllers are required, in accordance with wind-tunnel practice. At all times the velocity profile, including rotational flow, should be carefully measured by small and sensitive anemometer probes.

8.5 MEASUREMENTS OF WAVES

The wave is assumed to have been converted to an electrical quantity. For slow waves these may be recorded on a chart. The recorder itself may introduce errors and the taking of values from an inked line of finite thickness is possible only to limited accuracy. Alternatively, the signal may be converted to digital form by a digital meter, to a relatively large number of significant figures; of course, both the signal quantity and the sensor must "know" its value, and the time at which it is sampled, to an appropriate degree of precision. The availability of the signal in digital form allows the moments of a wave, or the harmonic content of a sinusoid, to be computed on a digital computer. For higher-frequency sine waves [say from one cycle per minute (0.1 rad sec^{-1}) upward] the measurement may be done by a nulling, i.e., a potentiometric, technique; this assumes that electrical signals of pure sinusoidal form are available with frequencies at least at the fundamental and second harmonic for reference. The method is similar in principle to that in a dc potentiometer, in that the signal to be measured is "backed off" by known signals; a detector indicates the success of complete backing off. For the alternating signal both the amplitude and the relative phase lag of the backing-off voltage must be adjusted, and this for every harmonic present. This has been found to be reasonably convenient above the frequency limit of 0.1 rad sec^{-1}.

The juggling of amplitudes and phase angles of the backing-off voltages could perhaps be eased by a slightly different arrangement using Lissajou figures; the signal is connected to the x plates of an oscillograph with long persistence, with the second, . . . , harmonic backing-off voltages in series. To the y plates is connected the fundamental backing-off voltage. The final figure is a straight line at 45° to the axis.

Whichever method is used, the amplitude and phase angle of the fundamental backing-off voltage needs to be known; the amplitude may be varied and measured in relative terms by a potential divider of high resolution, so that relative amplitudes may be measured to 0.1 %. The phase angle of the backing-off voltage is most conveniently and accurately measured by using a sine–cosine potentiometer, provided that a physical mounting is available which allows the phase angle to be adjusted and measured. Usually, commercially available potentiometers are not so made; a device was developed [9] that allows as many potentiometers as desired to be mounted so that each could be driven at its own speed (so generating fundamental and harmonic frequencies). Each could have its individual phase angle altered and measured to the order of a tenth of a degree by a micrometer. This rotated the plate holding the potentiometer by means of a peg engaged in a hole. A diagram, which also illustrates his improved method of measuring angles, is given by Goss [2].

REFERENCES

[1] G. A. Turner and M. J. Goss, A Method of Finding Simultaneously the Values of the Heat Transfer Coefficient, the Dispersion Coefficient and the Thermal Conductivity of the Packing in a Packed Bed of Spheres: Part III. Experimental Methods and Results. *AIChE J.* **17**, 592 (1971).

[2] M. J. Goss, Determination of Thermal Parameters of a Packed Bed of Spheres by a Method of High Precision Frequency Response. Ph.D. Thesis, Univ. of Waterloo (1969).

[3] J. Rawcliffe, Univ. of Manchester Inst. Sci. and Tech. Private communication.

[4] A. B. Kaufmann, Potentiometer Loading Errors. Radio-Electronic Engineering, p. 12 (September 1952).

[5] P. F. Deisler, Jr. and R. H. Wilhelm, Diffusion in Beds of Porous Solids. *Ind. Eng. Chem.* **45**, 1219 (1953).

[6] G. A. Turner, The Frequency Response of Some Illustrative Models of Porous Media. *Chem. Eng. Sci.* **10**, 14 (1959).

[7] H. Kramers and G. Alberda, Frequency Response Analysis of Continuous Flow Systems. *Chem. Eng. Sci.* **2**, 173 (1953).

[8] J. J. Keyes, Jr., Diffusional Film Characteristics in Turbulent Flow: Dynamic Response Method. *AIChE J.* **1**, 305 (1955).

[9] G. A. Turner and G. C. Dosser, to Fisons Fertilizers Limited, "A Device for Varying the Phase Angle between Standard Sinecosine Potentiometers." British Patent Appl. 50485/64.

Chapter 9

Practical Considerations and Examples

When the ideas in this book are to be applied to the measurement of quantities, then the account can be written either from the point of view of the experimenter making a choice of method or as a selection of accounts of determinations that have been made.

9.1 CONSIDERATIONS GOVERNING THE CHOICE OF EXPERIMENTAL METHOD

If it is assumed that the system is defined for him, then the experimenter has first the choice of the mathematical model that he thinks is both simple enough yet adequate enough, and second, the choice of pulse or wave. (He also has then the choice of experimental apparatus.)

Possibilities from which his second choice can be made are laid out in Table 9.1. Faced with this range, he has a decision to make that is to be based on what has been discussed in previous chapters. The relevant factors will be summarized below (the weighting and the choice are of course subjective).

9.1.1 Generation

The apparent ease of generating a pulse (or step) has led to its widespread use, but as has been stated, a true impulse is impossible to generate, while an approximation can lead to substantial disturbances in the flow. Furthermore, the resulting pulse has then either to be assumed to be an impulse, or it has to be measured. If measured, the effect of the probes themselves and their response to transients or higher harmonics need to be known. Furthermore, the large concentration change may give rise to nonlinear effects; the equation may not yield a convenient analytic solution or, if linearity is assumed, the distortion in the output pulse would give misleading results.

If a shaped pulse is required (for its Fourier transform), the above problems are compounded by additional practical difficulties in generating and measuring the required shape.

9.1.2 Measurement

The probes and associated meters do not have perfect responses; i.e., they themselves affect the signal, and thus, means of either knowing the magnitude of the effect or of allowing for it should be available. Again, the presence of longitudinal boundaries will affect the wave, and so, whether the probe is in the test section or out of it, the response will be that of a finite bed. This has been discussed in the relevant chapters, but the situation may be summarized thus when longitudinal boundaries are present:

Impulses (*no reservoir phases present*). Equations are available for $\bar{t}$ and σ^2 in Table 5.1 (but computation of D may be difficult). For approximations see Levenspiel and Bischoff [1].

Pulses (*no reservoir phases present*). Equations for moments may be found, but computation of D may be difficult.

Pulses (*reservoir phases present*). Equations for moments are difficult or impossible to obtain.

Sine Waves (*reservoir phases either present or absent*). Methods are available for computing the response, but in practice they are not needed; the effects of both longitudinal boundaries and of the measuring devices can be readily allowed for (Chapter 6).

TABLE 9.1

SUMMARY OF PRACTICAL ASPECTS OF DISTURBANCES CONSIDERED[a]

Disturbance	Shape at inlet: Description	Shape at inlet: Determination	Ease of generation	Information computed	Measurement	Can effect of (a) or (b) be readily found? (a) Probes	Can effect of (a) or (b) be readily found? (b) Longitudinal boundaries	Ease of simultaneous determination of several parameters
Impulse (step)	Dirac	Assumed	Impossible		Distribution	No	Yes[b]	Good[b]
Pulse	Specific (ramp, square, etc.)	Measured or assumed	Fair	Moments or distribution in Laplace domain or harmonic content or attributes of distribution	Distribution	No	No (?)	Fair (?)
	General	Measured	Good		Distribution	Yes	Yes	Good
					Isolated values	No	No	Zero
Wave	Periodic	Measured	Fair	Harmonic content (amplitude and phase)	Distribution	Yes	Yes	Good
Wave	Sine	Measured	Good	Amplitude and phase	Peak, trough, (intermediate for ψ)	Yes	Yes	Good
Random	Noise	(Not considered)	—	—	—	—	—	—

[a]Only measurements with time, and in the fluid, are considered.
[b]For simple cases.

9.1.3 Errors

It seems fair to generalize that the determination of the values of parameters involves small differences of large (measured) quantities, so even modest inaccuracies in measurement can give rise to highly inaccurate (or even ludicrous) answers. General considerations of accuracy have been discussed in the appropriate chapters, but some additional points may be made.

9.1.3.1 Moment Method

It will be remembered that the moment method is likely to be highly susceptible to errors because a more or less continuous distribution of values is required. So there will be both sampling errors and "noise" (uncertainty) and the effects of these are magnified because the readings with the greatest uncertainty—i.e., in the "tail"—are multiplied by the largest values of time, $(\text{time})^2$, etc.

An estimation of errors in calculated values of flow velocity and dispersion coefficient found from a simulated curve [calculated from Equation (2.37)] has been reported by Berger [2]. It was postulated that the "experimental" curve v_{ex} was given by

$$v_{ex} = v + (p_1 v/100) + (p_2 v_{max}/100)$$

where v is the value calculated from Equation (2.37). Thus, p_1 represents an error that is a fixed percentage of the instantaneous reading, while p_2 represents an error as a percentage of the maximum; i.e., it is constant for all time at a fixed value of z. It was also postulated that the curve would be truncated at some finite time such that $p_v = (v_{trunc}/v_{max}) \times 100$. Numerical values assumed were: $Q_{pl} = 1$ g mol cm^{-2}; $z = L = 15$ cm; $D = 45$ cm^2 sec^{-1}; $U = 45$ cm sec^{-1}; Pe $= 15$.

The simulation was done on an IBM 360/75 digital computer.

In Chapter 3 a number of equations were given, and several combinations of these equations can be used to find D, Pe, and U. For example:

(*a*) *Using* $t_{1/2}$

The value of U may be found from Section 3.3.2, while the value of Pe can be obtained from either Equation (3.12) (giving t_{peak}), or Equation (3.17) (giving $\bar{t}$), or Equation (3.25) (giving σ^2). Hence, the value of D can be found from the definition of Pe.

(*b*) *Using Moments*

Equations (3.12), (3.17), and (3.25) may be combined in three ways to obtain two equations, which can be solved simultaneously to produce values of U and D. In this example Equations (3.17) and (3.25) were so used, giving

$$D = (U/2)(\bar{t}U - L) \tag{9.1}$$

and

$$U = \{3\bar{t}L + [(3\bar{t}L)^2 - 4L^2(\sigma^2 - 2\bar{t}^2)]^{1/2}/2(2\bar{t}^2 - \sigma^2) \tag{9.2}$$

(*c*) *Using* t_{peak}

Equation (3.12), rewritten, is

$$\text{Pe} = (2Ut_{\text{peak}}/L)/[1 - (Ut_{\text{peak}}/L)^2] \qquad \text{(3.12) R}$$

while U can be found from Equation (9.2).

Results

It was found that, with the exception of $t_{1/2}$ (discussed below), the computations could be truncated at $t = 2$ sec without appreciable error.

(*a*) *Using* $t_{1/2}$

The best value of $t_{1/2}$ that could be calculated from the theoretical curve (viz., $p_1 = p_2 = 0$, $t_{\text{trunc}} = 2$ sec) instead of being the theoretical value of $L/U = 0.33$ sec, was in error by 3%. The values of D and Pe subsequently found were in appreciable error, as can be seen from Table 9.2, which lists

TABLE 9.2

COMPUTED VALUES FROM THEORETICAL CURVE[a]

Parameter	Ratio: computed to theoretical	Equation used
$t_{1/2}$	1.032	Section 3.3.2
U	0.969	Section 3.3.2
D	1.142	(3.12)
	0.712	(3.17)
	0.919	(3.25)
Pe	0.685	(3.12)
	1.361	(3.17)
	1.054	(3.25)
Other parameters		
t_{peak}	1.001	—
$\bar{t}$	1.000	(3.15)
σ^2	1.000	(3.20)
Q	0.999	(3.3)

[a] $p_1 = p_2 = 0$; $t_{\text{trunc}} = 2$ sec.

the equations used. The reason for this error is obscure, and values of $t_{1/2}$ were not computed for other values of p_1, p_2, or t_{trunc}.

It will be seen how large errors in D and Pe can be caused by a relatively small error in $t_{1/2}$.

(*b*) *Using Moments*

Effect of truncation on U

Very small (error $< 0.02\%$).

Effect of truncation on Pe *and D*

Results are given in Figure 9.1 as functions of p_v.

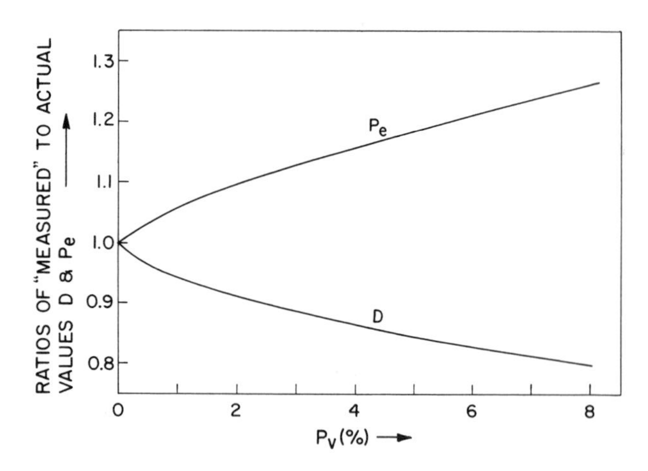

Fig. 9.1. Effect of "truncation" on "measured" values of Pe and D (moment method). (With acknowledgments to Berger [2].)

Effect of p_1

The first and second moments are unaffected by a constant percentage error; hence U, D, and Pe are unaffected.

Effect of p_2

The whole response curve is, in effect, shifted vertically so that in addition to the direct effect of the error in v on the first and second moments, there will be for all finite time limits a "truncation error"; that is to say, the computed value of the moments will depend upon the limits of the integration. The results from some illustrative examples are shown in Figure 9.2, which plots the ratio of calculated to actual values of D and Pe (computed from $\bar{t}$ and σ^2) as a function of the truncation (strictly, in terms of p_v computed from the actual "measured"—i.e., incorrect—curve). The values of p_2 chosen were 0.5 and 1.0; the sign of p_2 will make a difference, of course.

It will be noted that large percentage errors can arise in the "experimental" values.

(c) *Using Values at the Peak*

The magnitude of errors arising from a constant error in concentration, p_2, is less when the value of t_{peak} is found than when the moments are used. For, no matter what p_2 is, the peak time remains constant, while the value of U, although dependent on the value of $\bar{t}$ found, is not excessively dependent on either the value of p_2 or t_{trunc}. Thus, for the range of values of p_2 and t_{trunc} covered by Figure 9.2 the percentage error in D ranged from -2% to -20%, and the percentage error in Pe ranged from $+2\%$ to $+26\%$.

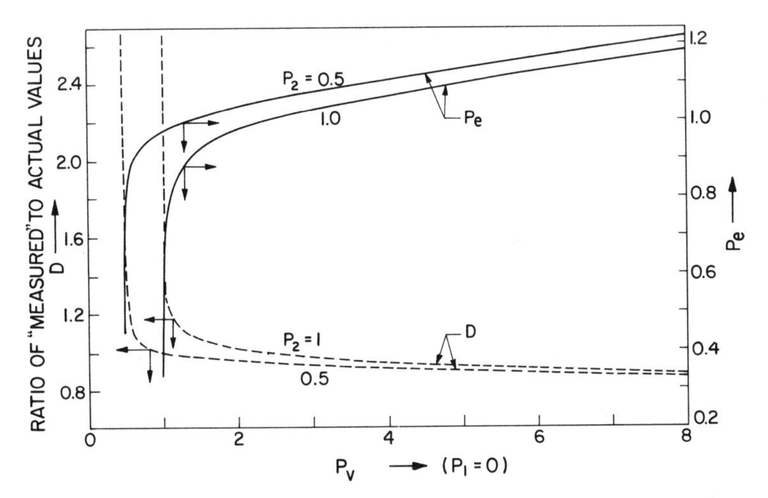

Fig. 9.2. Effect of "measurement error" on "measured" values of Pe and D (moment method). (With acknowledgments to Berger [2].)

Conclusions

Only a specific example has been given, so the results are not general. For this case it was found that large—even impossibly large—errors can arise from modest errors of measurement.

It is difficult to find $t_{1/2}$ precisely, and errors of a few per cent gave large errors in Pe and D.

Again, values of Pe and D computed from a simultaneous solution of the equations for the first and second moments were very sensitive to both a constant absolute error of concentration and to truncation error. On the other hand, a constant percentage error has no effect.

If the time at the peak of the wave were measured and combined with a value of U found from the first moment $\bar{t}$, the resulting values of Pe and D were very sensitive to small errors in U and t_{peak}. Both $\bar{t}$ and t_{peak} are prone to error, the former because it is a moment, the latter because it is difficult to find the exact time at the peak, especially if it is fairly flat.

The errors in moment methods is further discussed by Butt [3], Curl and McMillan [4], Anderssen and White [5], and Rony and Funk [6]. Again, in order that a measurable output signal be obtained, the peak value of a pulse must be large, being infinite for the limiting case of an impulse, and so nonlinear effects are likely to arise.

9.1.3.2 Frequency Response

In contrast to the above, a sine wave requires only values of the concentration at peak and trough to obtain the amplitude, which can be modest in size. The phase angle is more accurately obtained by concentration values near the midpoint of the swing. Chapter 8 points out the appreciable errors that can arise in phase-angle and amplitude measurements if unsuspected harmonics are present.

9.1.3.3 Modified Moment Method

The method suggested in Section 4.3 in which $(d^n/ds^n)[(1/F)\, dF/ds]$ is used, where F is the ratio of Laplace transforms of $v(t)$ in the pulse at two stations, although not requiring moments to be calculated, suffers from the risk of errors arising when a function is graphically differentiated. It involves measurement of $v(t)$, then computation of

$$\bar{v}(s) = \int_0^T v(t)e^{-st}\, dt/m_0$$

for a range of values of s; the resulting ratio of values of $\bar{v}(s)$ at inlet and outlet, viz., $F(s)$, is then differentiated, and finally the expression $(1/F)\, dF/ds$ is differentiated n times. These differentiation processes are likely to introduce appreciable errors. Furthermore, the range of values of s has to be carefully chosen.

9.2 ADVANTAGES OF SINE WAVES

In the light of the above difficulties and chances of error, it is recommended that sine waves be used. Their advantages may be summed up as follows:

1. Generation is easy.

2. The wave (in a linear system) is unchanged in form; this stems from the fact that all derivatives of a sine function are also sine functions.

3. The sine function is naturally smooth, and it and all derivatives are continuous.

4. From a given set of periodic continuous functions a set of normal, orthogonal periodic functions can be constructed, so any given periodic function can be expanded as the sum of a number of derived orthogonal functions. Thus, the response of a system may be regarded as that of the sum of a number of sinusoidal components; hence, the more closely the initial disturbance approaches a pure sine wave, the fewer the components that must be considered in the analysis.

5. Small amounts of heat or matter are needed per unit time. Temperature or concentration changes are relatively small, and the assumptions of both linearity and small heat loss through the walls will most likely be correct. Furthermore, parameter values are close to point values, rather than average values.

6. The system tends to be self-purifying; that is, the higher harmonics of an impure wave are attenuated faster than the fundamental, with consequent improvement in measurement accuracy as discussed in Chapter 8.

7. The sine wave can be represented completely by a line whose length represents its amplitude and whose orientation indicates its phase angle.

8. As the number of parameters in a system increases, there is available a variable, viz., the frequency, that can take any number of values without the wave shape being altered, and without the algebraic expression of response, for a fixed number of parameters, altering. Furthermore, as the number of parameters is increased, the algebraic expressions do not become unwieldy.

9. The effects that longitudinal boundary conditions and the measuring probes have on the wave are readily and rigorously allowed for.

10. Values of one, two, three, or more parameters can usually be found readily and accurately without curve-fitting and excessive computational time.

9.3 SUMMARY OF PARAMETERS FOUND AND METHODS USED

This refers to a defined system; i.e., one that can be described exactly by the linear equations. The problems of parameter identification, of comparison of models, is a broader subject not considered here.

Generally, the system may contain, and the model may postulate, M parameters. (A "parameter" may in fact be a group, containing several parameters, which can be found collectively. See, for instance, the remarks at the end of Example 2, Case 2, Section 4.2.2.) Of these M parameters, $M - N$ may be assumed to be found from other, independent, measurements, so in fact N parameters are now left to be determined. These N parameters may be of two classifications: (a) different kinds, (b) same kind, different numerical values. Thus, the general equation is

$$D\frac{\partial^2 v}{\partial z^2} - U\frac{\partial v}{\partial z} - \frac{\partial v}{\partial t} + \sum_{j=2}^{j} q_j = 0 \tag{9.3}$$

If the jth stationary source of flux contains r_j parameters ($\tilde{V}_j$ being perhaps included among them), then the most general case of Equation (9.3) will involve D, U, and all the others. That is,

$$M = 2 + \sum_{j=2}^{j} r_j$$

9.3.1 Residence-Time Distribution

A distribution of residence times is another example of information found by comparing the input and output signals of the system. It is an important topic, covered in many references, but is not considered here because (a) being a distribution, it has to be described either by an analytic expression or by moments, rather than by a parameter; (b) distributions of practical interest arise from actual systems, not from the simple wave propagating medium here discussed.

9.4 SOME GENERAL TREATMENTS

9.4.1 The Use of Sine Waves

(Note: The sign convention of the phase angle will be ignored in the following, i.e., numerical values only will be taken.)

The time response of a bed of length L, after the effects of longitudinal boundaries and of probes have been allowed for (as in Section 6.6), is obtained from Equations (6.26) and (6.28) for Equation (9.3) as

$$\ln [V^+(L)/V_s(0)] = \Pi = \{(U/2D) - [(s^* + b^*)/2]^{1/2}\}L \tag{6.26 R}$$

$$|\psi(L)| = [(s^* - b^*)/2]^{1/2}L \qquad \text{if} \quad \psi(0) = 0 \tag{6.28 R}$$

where the asterisk has been inserted to allow for the more general case where there are j stationary sources of flux in parallel, each having its own value of the complex shunt admittance (Chapter 7). In other words, Equations (6.25),

(6.23), and (6.24) become, respectively,

$$s^* = [(b^*)^2 + (g^*)^2]^{1/2} \tag{9.4}$$

$$b^* = (U/2D)^2 + (1/D) \sum_{j=2}^{j} \tilde{V}_j Y_{1j} \tag{9.5}$$

$$g^* = (1/D)\left(\sum_{j=2}^{j} \tilde{V}_j Y_{2j} + \omega\right) \tag{9.6}$$

If it is remembered that the numerical values of Π and ψ are known from the experimental measurements, then the value of a new variable ν $(= \Pi^2 - \psi^2)$ can be found at any frequency. Thus, from Equations (6.26), (6.28), and (9.4)–(9.6) it can be seen that

$$\frac{\nu(\omega)}{\text{Pe}} - \frac{L}{U} \sum_{j=2}^{j} \tilde{V}_j Y_{1j}(\omega) = \Pi(\omega) \tag{9.7}$$

If there are N parameters, then measurements at N frequencies give N linear first-order equations all of the same form as Equation (9.7). In these, only ν and Π are known. However, a trial value of Y_{ij} (obtained from trial values of the parameters in its makeup) at one frequency fixes the values at all other frequencies, so that these, becoming "coefficients" in the equation, are thus temporarily removed as unknowns. The test of the correctness of choice of the values of the appropriate parameters rests on the strong condition that the eliminant (the terms of which will be made up of these "coefficients" along with other known quantities) of $(n + 1)$ linear nonhomogeneous equations in n unknowns must be zero, and by virtue of the above temporary transmutation, n will be less than N. The value of neither U, L, nor Pe necessarily needs to be known; in fact, the value of Pe and $L\tilde{V}/U$ can be found.

The procedure is best illustrated by some examples.

(*i*) ***No Reservoir Phase; D ≠ 0*** (Case b of Section 6.3.2)

The general equation is

$$D\frac{\partial^2 v}{\partial z^2} - U\frac{\partial v}{\partial z} - \frac{\partial v}{\partial t} = 0 \tag{1.5b) R}$$

Thus, if $\tilde{V}_j$ is put equal to zero, Equation (9.7) becomes

$$\nu/\Pi = \text{Pe} \tag{9.8}$$

for all ω, so Pe can be found.

Again, Equations (6.32) and (6.34) [along with (6.30)] can be arranged as

$$\Pi = \frac{\text{Pe}}{2}\left[1 - \left(\frac{\{1 + 4[N_f/(\text{Pe}/2)]^2\}^{1/2} + 1}{2}\right)^{1/2}\right] \tag{9.9}$$

$$|\psi| = \frac{\text{Pe}}{2}\left(\frac{\{1 + 4[N_f/(\text{Pe}/2)]^2\}^{1/2} - 1}{2}\right)^{1/2} \tag{9.10}$$

from which N_f (frequency number [7] $= L\omega/U$) can be found. That is, U can be found if ω and L are known.

(ii) No Reservoir Phase, but D = 0

There is no amplitude attenuation, but

$$|\psi| = N_f = L\omega/U \tag{9.11}$$

so again U can be found.

(iii) Reservoir Phases Present, but D = 0 (Case c of Section 6.3.2)

This is the situation first discussed by Rosen and Winsche [8]. See Equations (6.38) *et seq.* Thus,

$$\Pi = -Y_1\tilde{V}L/U \tag{9.12}$$

which is also obtained by putting Pe $= \infty$ into Equation (9.7) and

$$|\psi| = (Y_2\tilde{V} + \omega)L/U \tag{9.13}$$

Values of Y_1 and Y_2 may thus be found if the other quantities in the equation are known. The finding of parameters making up Y is considered in Section (iv).

(iv) Reservoir Phases Present, but D ≠ 0

This is the most general case in the system described. The equation will be

$$D\frac{\partial^2 v}{\partial z^2} - U\frac{\partial v}{\partial z} - \frac{\partial v}{\partial t} + \sum_2^j q_j = 0 \qquad \text{(9.3) R}$$

The total number of parameters will be $D + U + (j - 1) \times$ number of parameters associated with q_j (see Section 9.3). It is perhaps convenient to divide the possibilities into two cases.

Case (i)

$U =$ unknown; $D =$ unknown; $j = 2$ (i.e., one reservoir phase) with r parameters, perhaps including an unknown $\tilde{V}_2$ among them. Thus, the total number of parameters would be $2 + r$.

Case (ii)

$U =$ known; $D =$ known; $j = 2, 3, \ldots, j$; $r = 1$ in the makeup of each admittance; in addition, $\tilde{V}_j$ is unknown. Thus the total number of parameters would be $2(j - 1)$. [It is possible to combine Cases (i) and (ii) and to make $r > 1$ in Case (ii).]. These cases are now examined.

Case (i)

(a) If $r = 1, 2, 3, \ldots, r$, then Equation (9.7) becomes

$$[\nu(\omega)/\mathrm{Pe}] - (L/U)\tilde{V}_2 Y_1(\omega) = \Pi(\omega) \tag{9.14}$$

Now, the value of $Y_1(\omega)$ is unknown, but if a trial value of all parameters $\pi_1, \pi_2, \ldots, \pi_r$ involved in its makeup are chosen, the value of $Y_1(\omega)$ is then specified for that value. So, for these trial values we can write the equations for three values of the frequency as

$$[\nu(\omega_1)/\mathrm{Pe}] - (L/U)\tilde{V}_2 Y_1(\omega_1) = \Pi(\omega_1) \tag{9.15a}$$

$$[\nu(\omega_2)/\mathrm{Pe}] - (L/U)\tilde{V}_2 Y_1(\omega_2) = \Pi(\omega_2) \tag{9.15b}$$

$$[\nu(\omega_3)/\mathrm{Pe}] - (L/U)\tilde{V}_2 Y_1(\omega_3) = \Pi(\omega_3) \tag{9.15c}$$

These can be looked upon as three linear nonhomogeneous algebraic equation in two unknowns, viz., 1/Pe and $L\tilde{V}_2/U$. For these to be simultaneously satisfied, the determinant of the coefficients must be zero. That is,

$$\begin{vmatrix} \nu(\omega_1) & Y_1(\omega_1, \pi_1, \pi_2, \ldots) & \Pi(\omega_1) \\ \nu(\omega_2) & Y_1(\omega_2, \pi_1, \pi_2, \ldots) & \Pi(\omega_2) \\ \nu(\omega_3) & Y_1(\omega_3, \pi_1, \pi_2, \ldots) & \Pi(\omega_3) \end{vmatrix} = 0$$

or, rewritten,

$$f_1(\pi_1, \pi_2, \ldots, \pi_r) \equiv K_1(Y_{11}/Y_{13}) + K_2(Y_{12}/Y_{13}) + 1 = 0 \tag{9.16}$$

where

$$K_1 = (\Pi_3\nu_2 - \Pi_2\nu_3)/(\Pi_2\nu_1 - \Pi_1\nu_2) \tag{9.17}$$

$$K_2 = (\Pi_1\nu_3 - \Pi_3\nu_1)/(\Pi_2\nu_1 - \Pi_1\nu_2) \tag{9.18}$$

and the second or only subscript refers to the frequency.

Thus, trial values of the parameters $\pi_1, \pi_2, \ldots, \pi_r$ are found from which the computed values of $Y_1(\omega, \pi_1, \pi_2, \ldots, \pi_r)$ make f_1 equal to zero. Function f_1 and the parameters are in the $(f_1, \pi_1, \pi_2, \ldots, \pi_r)$ space. Thus, if there is only one parameter, it is as in Figure 9.3a. Again, if there are two parameters, the space is as shown in Figure 9.3b, but instead of a unique combination $\pi_1{}^*$, $\pi_2{}^*$ being found, a continuous distribution is obtained and the state $f_1 = 0$ is a line a in Figure 9.3c. However, if a fourth frequency is used, giving rise to another equation in the set (9.15), then by taking another combination of three of these equations at a time a second line b in Figure 9.3(c) is obtained whose intersection with line a gives $\pi_1{}^*$ and $\pi_2{}^*$, the desired values. (Additional frequencies can be taken as check values; the closeness of the additional intersections will depend on the size of experimental errors.)

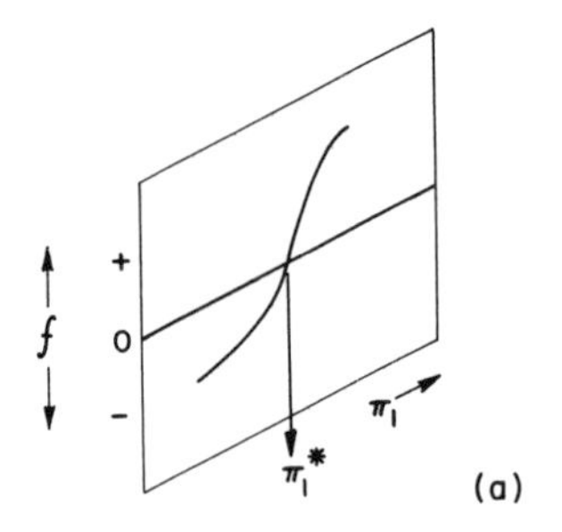

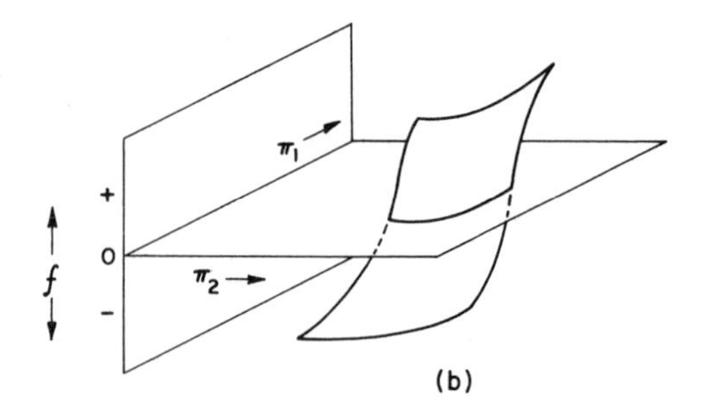

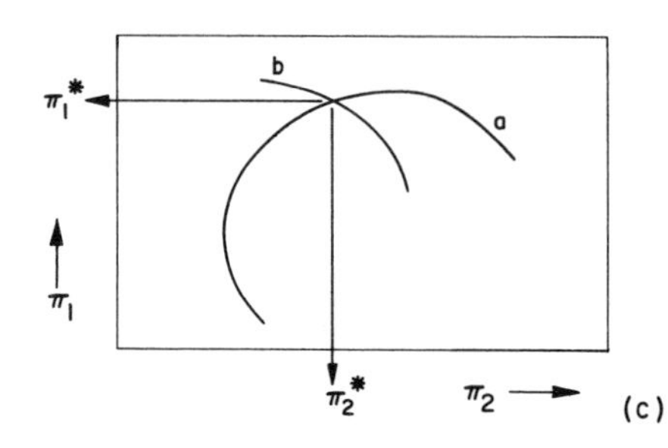

Fig. 9.3. (a) Space diagram for (f_1, π_1) line; admittance a function of one parameter, π_1. (b) Space diagram for (f_1, π_1, π_2) surface; admittance a function of two parameters, π_1 and π_2. (c) The curves of $f_1 = 0$. Curves *a* and *b* from different combinations of three frequencies taken from four.

Presumably, 3, 4, . . . parameters (in addition, of course, to Pe and U) could be found in this way by taking the appropriate number of frequencies, and working in 4-, 5-, . . . dimensional space.

The values of the "coefficients" and the "constant terms" in Equations (9.15) being now known, from any two equations one of the "variables," viz., 1/Pe or $L\tilde{V}_2/U$, can be eliminated, to give an equation for the other that can now be solved; viz.,

$$\frac{1}{\text{Pe}} = \begin{vmatrix} \Pi_1 & Y_{11} \\ \Pi_2 & Y_{12} \end{vmatrix} \Bigg/ \begin{vmatrix} \nu_1 & Y_{11} \\ \nu_2 & Y_{12} \end{vmatrix} \tag{9.19}$$

and

$$\frac{L\tilde{V}}{U} = \begin{vmatrix} \nu_1 & \Pi_1 \\ \nu_2 & \Pi_2 \end{vmatrix} \Bigg/ \begin{vmatrix} \nu_1 & Y_{11} \\ \nu_2 & Y_{12} \end{vmatrix} \tag{9.20}$$

Thus, for example, the model might be of a porous medium, with relative volume $\tilde{V}$ of dead space of specified geometry defined by one parameter π_1 (e.g., depth of a cylindrical pore in communication with Phase 1 by transport processes but not by any hydrodynamic flow). So, from measurements at three frequencies it will be possible to find the values of this depth, the longitudinal Peclet number, and of $L\tilde{V}/U$.

Again the model might be of heat transfer in a packed bed of solids. The admittance Y_1 can now be a function of two parameters, viz., $\pi_1 \equiv$ conductivity of the solid $a\rho C$, and $\pi_2 \equiv$ heat transfer coefficient h. That is, there are four unknowns, viz., Pe, $U\tilde{V}/L$, $a\rho C$, and h, and, as stated, four values of the frequency are needed [9–12].

The number of parameters can be increased; workers in the fields of chromatography and catalysis tend to invoke many parameters because of the complexity of the mechanism (see, e.g., Section 7.5). The present method does not appear to have been used, but an example, treated in two different ways, is discussed briefly in Sections 9.4.2 and 9.4.3. It involves a number of parameters. (Simpler models can be thought of as special cases of the more general ones, but oversimplification, although making for ease of determination, can give misleading results. See, for example, the discussion by Rony and Funk [6].)

As pointed out [9–12], the finding of $\pi_1{}^*$, $\pi_2{}^*$ is quick and easy. However, a watch must be kept on the values of Y_1; in certain ranges of its domain its behavior may be such that the method breaks down.

Case (ii) (General)

If the differential equation is

$$D\frac{\partial^2 v}{\partial z^2} - U\frac{\partial v}{\partial z} - \frac{\partial v}{\partial t} + \sum_{j=2}^{j} q_j = 0$$

and $q_j = \tilde{V}_j Y_j v$ [from Equation (7.3a)], then Equation (9.14) becomes

$$[\nu(\omega)/\text{Pe}] - (L/U)[\tilde{V}_2 Y_{12}(\omega) + \tilde{V}_3 Y_{13}(\omega) + \cdots] = \Pi(\omega) \qquad (9.21)$$

This is an extension of the above; there are now $2 + (j-1) + (j-1)n$ unknowns, viz.; Pe; L/U; $\tilde{V}_2, \tilde{V}_3, \ldots, \tilde{V}_j$; $Y_{12}, Y_{13}, \ldots, Y_{1j}$, if Y_{1j} is a function of $n\ (= r_j)$ parameters.

Case (ii) (Limited)

In principle the general case could be tackled along the same lines, but it is perhaps doubtful whether reliable information could be obtained from so general a model, and it might be desirable to simplify it, say by making $N < M$. Thus, in one use of the method [13, 14], viz., that of finding the distribution of sizes of dead spaces in a porous bed, the values of Pe and L/U were obtained by other means, so that the equation becomes

$$\tilde{V}_2 Y_{12}(\omega) + \cdots + \tilde{V}_j Y_{1j}(\omega) = [\Pi - (\nu/\text{Pe})]U/L = \mu(\omega) \qquad (9.22)$$

the value of the right-hand side ("the constant term" μ) being known at any frequency.

If again $\tilde{V}$ is treated as the "unknown," the "coefficients" Y, also unknown, are dependent on the value of the parameter or parameters that they contain.

So the value(s) of these parameter(s) must be found to make

$$f_3(\pi_1, \pi_2, \ldots, \pi_n) \equiv \begin{vmatrix} Y_{121} & Y_{131} & \cdots & Y_{1j1} & \mu_1 \\ \vdots & \vdots & & \vdots & \vdots \\ Y_{12j} & Y_{13j} & \cdots & Y_{1jj} & \mu_j \end{vmatrix} = 0 \qquad (9.23)$$

If, say, Y contains one parameter, it is first necessary to try $j - 1$ values simultaneously, the combination making $f_3 = 0$. This may not be the only combination to do so, but it will be recalled that there are [for this case, viz., one parameter, π_1, that can have any numerical value denoted by $(\pi_1)_j$], $2(j - 1)$ unknowns, viz., $\tilde{V}_2, \ldots, \tilde{V}_j$ and $(\pi_1)_2, \ldots, (\pi_1)_j$. So, if measurements are made at $2(j - 1)$ frequencies, it will be possible to take $j - 1$ independent eliminants, all of which must be made simultaneously equal to zero by the chosen values of $(\pi_1)_2, \ldots, (\pi_1)_j$.

EXAMPLES

This has been done by Turner [13, 14] for experimental models of a packed bed: in Model 1 the value $\tilde{V}_j$ was of dead space of length l_j; the dead space was of pockets or blind pores that communicated with the channels through which fluid could flow. Material could diffuse in and out of the blind pores but fluid could not flow through them. Hence, $\pi_1 \equiv l$.

Model 2 had no blind pores but consisted of circular capillaries having a distribution of lengths such that there are n_{ji} channels in the porous material, each capillary being of length l_j and radius r_i. Thus, the volume of these channels will be $\tilde{V}_{ji} = n_{ji}\pi r_i^2 l_j$.

The physical situation in Model 2 is a little different from that in Model 1 inasmuch as there is no reservoir phase, but the same kind of simultaneous equations result. For, if bold type represents the resulting (measured) amplitude and phase angle of the sine wave at the exit from the bed, it will be the sum of the coherent interfering waves issuing from all the capillaries, as Appendix 10 explains:

$$\mathbf{V} \cos \boldsymbol{\psi} = (1/F) \sum^{j} \sum^{i} n_{ji} F_{ji} V_{ji} \cos \psi_{ji} \qquad (9.24)$$

and

$$\mathbf{V} \sin \boldsymbol{\psi} = (1/F) \sum^{j} \sum^{i} n_{ji} F_{ji} \sin \psi_{ji} \qquad (9.25)$$

where $F = \sum^{j} \sum^{i} n_{ji} F_{ji}$ is the total flow rate (cm^3 sec^{-1}); F_{ji} is the flow rate in the jith capillary (cm^3 sec^{-1}); and V_{ji} and ψ_{ji} are the amplitude and phase angle, respectively, of the wave issuing from the jith capillary.

The magnitudes V_{ji} and ψ_{ji} will depend on the process going on inside the capillary. That is, on whether there is diffusion or not, on the velocity profile (and on the cross-sectional shape, if not circular), and expressions for the measured quantities Π and ψ (and so ν) can be derived in terms of the last-mentioned quantities together with the unknowns $\tilde{V}_{ji}$, r_i, l_j. Thus there arises an equation of type similar to Equation (9.22). It is

$$\sum\sum \tilde{V}_{ji}\eta_{ji}(\omega) = \mu(\omega) \tag{9.26}$$

where η_{ji} is a known analytic expression involving the unknown quantities r_i and l_j. The trial procedure is to choose a value of r_i, and for each of these values choose values of l_j. For each combination the value of η_{ji} is calculated at a certain value of ω, and the condition to be satisfied is that the eliminant is zero, as before.

As postulated, the procedure is designed to give a bivariate distribution of $\tilde{V}_{ji}$ against r_i and l_i. If the optimization process of making the determinant zero is too long or too uncertain, then the problem can be reduced in magnitude by finding a univariate distribution that approximates to (at least the frequency response of) the actual system. In other words, a mean value of either r_i or l_j is found and the distribution of $\tilde{V}_i$ against r_i or $\tilde{V}_j$ against l_j is found. That is, Equation (9.26) becomes

$$\sum \tilde{V}_i\eta_i(\omega) = \mu(\omega) \tag{9.27a}$$

or

$$\sum \tilde{V}_j\eta_j(\omega) = \mu(\omega) \tag{9.27b}$$

respectively, both of which are of the same form as Equation (9.22). As Aris [15], points out, Models 1 and 2 may be combined, and either model can be extended to cover the case of continuous (rather than discrete) distributions of parameter (viz., length or diameter).

Singh [16] has used a simulated Model 1, from which Equation (9.22) involved six unknowns, as well as a simulated model involving ten unknowns. It appears to be possible to reclaim the simulated values. No examples of the method are known in which Y or η involved more than one parameter.

9.4.2 Curve-Fitting (Optimization) Technique

This process consists of trying different values of the wanted parameters until quantities (such as those listed below) computed from the chosen equation (i.e., the model) agree with the corresponding values as measured experimentally. It is covered in the literature and in texts; for parameter determination, the quantity that suits the problem best is chosen. The most likely ones are as follows.

(a) Amplitudes or phase angles (or both) of a steady state sinusoidal concentration wave, or of the harmonic components of a steady cyclic wave, or of the harmonic components of a pulse. The investigation is said to be in the *frequency domain.*

(b) The values of concentration at various times during transit of a concentration pulse. The investigation is now said to be in the *time domain.*

(c) The values of the Laplace-transformed variable, computed for various values of the transformation variable s. Correspondingly, this is said to be in the *Laplace domain.*

(d) The last of the more widely used quantities are the n moments, m_n, of a concentration pulse.

In all of these cases, except the time domain, the concentration $v(z, t)$ is operated upon by the process:

$$\text{transformed variable} = \int_0^\infty v(z, t) \times (T)\, dt$$

where $T \equiv e^{i\omega t}$ for the Fourier transform for work in the frequency domain; its attributes were discussed in Chapter 4; or $T \equiv e^{-st}$ for the Laplace transform for work in the Laplace domain; it has the useful feature that its effect lessens as t increases, that is as the concentration in a pulse becomes smaller and more uncertain; or $T \equiv t^n$ for the moments; as stated earlier, the effect increases as t increases, especially for larger values of n, and so the uncertainty in the tail of a pulse is magnified. A few examples of reported work are now given.

Frequency Domain

In practice a search is made to make zero or a minimum the square of the differences between measured and computed values. Thus, a one-, two-, . . . , N-dimensional search has to be performed using the amplitude or phase angle at one, two, . . . , N values of the frequency. This may involve large amounts of computer time and more than one minimum or zero may exist. Gangwal *et al.* [17], after a five-dimensional search using either amplitude ratios or phase angles (the result of the Fourier transform of an assumed impulse) obtained values of a_2, D, $\mathbf{K}_2$, k_1, and k_2, from Kučera's model (Chapter 7).

Time Domain

Coats and Smith [18] used the analytic solution of a step change of concentration in a semiinfinite system in which there were dead pore spaces of relative volume $\tilde{V}$, of one fixed dimension, and having an infinite ratio of diffusivity to depth, as in Lapidus and Amundson [19] and Chao and Hoelscher [20]. By curve-fitting, some values of $\tilde{V}$, Pe, and the group kL/U (in which k was a transfer coefficient between channel and dead pore space) were found by minimizing the difference between the measured and theoretical response. A simpler example was discussed at the beginning of Chapter 3, but the fitting was in a "*distance domain,*" i.e., in z at constant t.

Laplace Domain

Veeraraghavan and Silveston [21] compared computations in the Laplace, frequency, and moment domains in an investigation of residence-time distributions.

Moments

Expressions for moments have been derived for relatively complicated systems; these expressions will contain the parameters of the model and it should be possible to extract values of these parameters by considering the first, second, . . . , moments simultaneously. However, the complexity of the algebra when a number of parameters are involved, and the errors of the computed moments, often make it necessary or desirable to simplify the assumptions and treatment. The Kučera–Kubin model has been referred to in Chapter 7. Schneider and Smith [22], for example, used the differences of first moments of a pulse to find the thermodynamic equilibrium constant $\mathbf{K}$, the only factor involved apart from density and relative volumes. On the other hand, the expression for the second moment is much more complicated and separate experiments and an assumption were introduced so as to reduce the number of parameters to be found. That is, $N < M$ (see Section 9.3).

9.4.3 Exact Solution

This is, of course, a desirable method but is possible only for cases where one or two parameters are involved; for example, U and Pe, as has been discussed in many places in this book. Occasionally it is possible to do this for rather more complicated systems incorporating a reservoir phase if approximations and assumptions are made. Thus, Keyes [23] has used amplitude ratios to calculate directly the thickness of the boundary layer when turbulent flow exists in a circular pipe, but several (probably justifiable) assumptions were made, including one that the velocity of the boundary layer was not significant.

Deisler and Wilhelm [7] also used amplitude ratios to find the longitudinal dispersion coefficient and the diffusion coefficient into a solid; their calculation likewise involved approximations and assumptions.

The method of Østergaard and Michelsen has been referred to in Chapter

4, as Equation (4.36). In that example no reservoir phase existed and no approximations were needed to be made.

9.5 WAVES IN THE RESERVOIR PHASE

In practice, the measurements of concentration used for parameter determination are almost invariably made in the flowing phase 1. In principle, they could alternatively be made in a reservoir phase, especially if this were a gas or liquid, but as will be realized from the concepts in Chapter 7, there can be a train of waves in a reservoir phase in directions normal to z. If to overcome this, area-average concentrations $\bar{v}_j$ in the reservoir (normal to z) are used, it can be readily deduced that there is a longitudinal wave of $\bar{v}_j$; this is interesting in that, unlike most other kinds of waves, there need be no direct mechanism by which the wave is transmitted longitudinally. For example, the longitudinal diffusivity in the reservoir may be—and usually is—zero.

For steady-state sinusoidal waves, viz., $\bar{v}_j = \bar{V}_j e^{i\omega t}$, it will be found that by using Equations (7.1b) and (7.3d) to eliminate v_1 and q_j from Equation (1.5a) (i.e., when $D \neq 0$), the ordinary differential equation in $\bar{V}_j$ that results is the same as Equation (6.18) for V_1. Again, when $D = 0$ the resulting ordinary differential equation for $\bar{V}_j$ is the same as Equation (6.38) for V_1. Hence the solutions will be the same as Equations (6.19), (6.20) *et seq.*, and Equation (6.39), respectively.

There will, of course, for both the above cases be a difference between the amplitudes and phase angle of the waves in phase 1 and phase j, and these will depend upon the boundary conditions. For the special case of the positive wave (i.e., an infinite bed) the relation between the complex amplitudes in the two phases (for both $D \neq 0$ and for $D = 0$) is $\mathfrak{v}_j{}^+ = (Y_2 - iY_1)\mathfrak{v}_1{}^+/\omega \equiv (|Y|\mathfrak{v}_1{}^+/\omega)e^{i(3\pi/2+\phi)}$ where $|Y| = (Y_1{}^2 + Y_2{}^2)^{1/2}$ and $\phi = \tan^{-1} Y_2/Y_1$. Thus, a concentration

$$v_1{}^+(z, t) = \mathfrak{v}_1{}^+(z) \cos[\omega t - \psi(z)]$$

in phase 1 would give rise to a wave

$$\bar{v}_j{}^+(z, t) = \bar{\mathfrak{v}}_j{}^+(z) \sin[\omega t - \psi(z) + \phi]$$

in the jth reservoir phase, the relation between the amplitudes being as just given.

When, on the other hand, there actually is longitudinal diffusion in the reservoir phase, the expressions for the steady sinusoidal waves in either phase 1 or phase j are not made up merely of two parts [as in Equation (1.10a)] but, a fourth-order ordinary differential equation arising, they will be comprised of four parts; that is V_1 and $\bar{V}_j$ are both of the form

$$V_p = \sum_{i=1}^{4} A_{pi} e^{\gamma_{pi} z}, \qquad p = 1, j \tag{9.28}$$

Thus, for example, Littman, Barile and Pulsifer [24] used Equation (1.5a) with the Danckwerts boundary condition, and Equation (1.5e) (with a flux term q_j added to couple the equations together) along with the conditions of zero flux at the beginning and end of the stationary phase. Their resulting equations were of the form of Equation (9.28).

ADDITIONAL REFERENCES

Even in the case of plane kinematic waves in one-dimensional flow, considered here, the supplementary papers and books are legion. In particular, each branch of knowledge to which the treatment and methods can be applied will have its own rather specialized literature. Rather than an attempt to list any of these, the suggestion is made that the following few books or references will be found relevant.

On the subject of waves, there are many texts, some general, some applying to a specific aspect. Moore [25] is a clear account of waves in electrical and other systems, both dispersive and nondispersive, while Sharman [26], for example, is more general.

For a practical account of manipulations and computations involving sine waves, Manley [27] is very good.

The general problems in setting up mathematical models is covered by Himmelblau and Bischoff [28]. There have been several appraisals of models of the kind of wave equation discussed in this book. Asbjørnsen and Wang [29] list these and discuss modified forms of Equation (1.5a) with the Danckwerts boundary conditions and of Equation (1.5c), both for the steady sinusoidal state. They stress the large effect that ignoring longitudinal dispersion can have on the values of parameters as computed.

Applications of Fourier transforms are described in a practical manner by Jennison [30] and Stuart [31].

As will be appreciated, the success of the use of kinematic waves where there are reservoir phases will hinge on the choice of the laws governing transport between these and the flowing fluid. Many have been put forward, especially in mass transfer; some are tractable, others less so, and these would have to be linearized to allow of the kind of analytical treatment used above. Beveridge [32] has produced an extremely useful summary of many of these laws, while the review of Barker [33, 34] contains many references in heat transfer.

REFERENCES

[1] O. Levenspiel and K. B. Bischoff, Patterns of Flow in Chemical Process Vessels, "Advances in Chemical Engineering" (T. B. Hoopes, J. W. Drew, and T. Vermeulen, eds.), Vol. 4. Academic Press, New York, 1963.
[2] D. Berger, Private communication, Univ. of Waterloo.
[3] J. B. Butt, A Note on the Method of Moments. *AIChE J.* **8**, 553 (1962).
[4] R. L. Curl and M. L. McMillan, Accuracy in Residence Time Measurements. *AIChE J.* **12**, 819 (1966).
[5] A. S. Anderssen and E. T. White, The Analysis of Residence Time Distribution Measurements using Laguerre Functions. *Canad. J. Chem. Eng.* **47**, 288 (1969).
[6] P. R. Rony and J. E. Funk, Retention Time and First Time Moment in Elution Chromatography: General Conclusions. *Int. Symp. Advan. Chromatogr. Miami, 6th*, June 2–5 (1970).
[7] P. F. Deisler, Jr. and R. H. Wilhelm, Diffusion in Beds of Porous Solids. *Ind. Eng. Chem.* **45**, 1219 (1953).
[8] J. B. Rosen and W. E. Winsche. The Admittance Concept in the Kinetics of Chromatography. *J. Chem. Phys.* **18**, 1587 (1950).
[9] G. A. Turner, A Method of Finding Simultaneously the Values of the Heat Transfer Coefficient, the Dispersion Coefficient, and the Thermal Conductivity of the Packing in a Packed Bed of Spheres: Part 1. Mathematical Analysis. *AIChE J.* **13**, 678 (1967).
[10] M. J. Goss, Determination of Thermal Parameters of a Packed Bed of Spheres by a Method of High Precision Frequency Response. Ph.D. Thesis, Univ. of Waterloo (1969).
[11] M. J. Goss and G. A. Turner, A Method of Finding Simultaneously the Values of the Heat Transfer Coefficient, the Dispersion Coefficient, and the Thermal Conductivity of the Packing in a Packed Bed of Spheres: Part II. The Technique of Computing the Numerical Values. *AIChE J.* **17**, 590 (1971).
[12] M. J. Goss and G. A. Turner, idem, Part III. Experimental Method and Results. *AIChE J.* **17**, 592 (1971).
[13] G. A. Turner, The Flow Structure in Packed Beds. *Chem. Eng. Sci.* **7**, 156 (1958).
[14] G. A. Turner, the Frequency Response of Some Illustrative Models of Porous Media. *Chem. Eng. Sci.* **10**, 14 (1959).
[15] R. Aris, Diffusion and Reaction in Flow Systems of Turner's Structures. *Chem. Eng. Sci.* **10**, 80 (1959).
[16] P. Singh, Private communication, Univ. of Waterloo.
[17] S. K. Gangwal, R. R. Hudgins, A. W. Bryson, and P. L. Silveston, Interpretation of Chromatographic Peaks by Fourier Analysis. *Can. J. Chem. Eng.* **49**, 113 (1971).
[18] K. H. Coats and B. D. Smith, Dead-End Pore Volume and Dispersion in Porous Media. *Soc. Pet. Eng. J.* p. 73 (March 1964).
[19] L. Lapidus and N. R. Amundson, Mathematics of Adsorption in Beds. VI. The Effect of Longitudinal Diffusion in Ion Exchange and Chromatographic Columns. *J. Phys. Chem.* **56**, 984 (1952).
[20] R. Chao and H. E. Hoelscher, Simultaneous Axial Dispersion and Adsorption in a Packed Bed. *AIChE J.* **12**, 271 (1966).

[21] S. Veeraraghavan and P. L. Silveston, Residence Time Distributions in Short Tubular Vessels. *Canad. J. Chem. Eng.* **49**, 346 (1971).
[22] P. J. Schneider and J. M. Smith, Adsorption Rate Constants from Chromatography. *AIChE J.* **14**, 762 (1968).
[23] J. J. Keyes, Diffusional Film Characteristics in Turbulent Flow: Dynamic Response Method. *AIChE J.* **1**, 305 (1955).
[24] H. Littman, R. G. Barile, and A. H. Pulsifer, Gas-Particle Heat Transfer Coefficients in Packed Beds at Low Reynolds Numbers. *Ind. Eng. Chem. Fundam.* **7**, 554 (1968).
[25] R. K. Moore, "Traveling Wave Engineering." McGraw-Hill, New York, 1960.
[26] R. V. Sharman, "Vibrations and Waves." Butterworths, London and Washington, D.C., 1963.
[27] R. G. Manley, "Waveform Analyses." Chapman & Hall, London, 1950.
[28] D. M. Himmelblau and K. B. Bischoff, "Process Analysis and Simulation: Deterministic Systems." Wiley, New York, 1968.
[29] O. A. Asbjørnsen and B. Wang, Heat Transfer and Diffusion in Fixed Beds. *Chem. Eng. Sci.* **26**, 585 (1971).
[30] R. C. Jennison, "Fourier Transforms and Convolutions for the Experimentalist." Pergamon Press, New York, 1961.
[31] R. D. Stuart, "An Introduction to Fourier Analysis." Methuen, London, 1966.
[32] G. S. G. Beveridge, A Survey of Interphase Reaction and Exchange. Harkness Fellowship Report, Dept. of Chem. Eng., Univ. of Minnesota, 1962.
[33] J. J. Barker, *Ind. Eng. Chem.* **57** (4), 43 (1965).
[34] J. J. Barker, *Ind. Eng. Chem.* **57** (5), 33 (1965).

Appendix 1

Change of Variable in Moving Systems: Time*

Let

$$C = f(z, t) \tag{A1.1}$$

and suppose

$$t = \phi(w, x) \tag{A1.2}$$

Then if, say, z is kept constant, one can think of the situation as Equation (A1.1) plus the fact that $t = f(w)$. The total derivative (Edwards [2]) is then

$$\left(\frac{\partial C}{\partial w}\right)_x = \left(\frac{\partial C}{\partial z}\right)_t \left(\frac{\partial z}{\partial w}\right)_x + \left(\frac{\partial C}{\partial t}\right)_z \left(\frac{\partial t}{\partial w}\right)_x \tag{A1.3}$$

*See, for example, Edwards [1].

Similarly, if w is kept constant,

$$\left(\frac{\partial C}{\partial x}\right)_w = \left(\frac{\partial C}{\partial z}\right)_t \left(\frac{\partial z}{\partial x}\right)_w + \left(\frac{\partial C}{\partial t}\right)_z \left(\frac{\partial t}{\partial x}\right)_w \tag{A1.4}$$

If one desires to measure times from an origin that is of magnitude (distance/velocity) later than the fixed origin $t = 0$, then if θ denotes times measured from this new origin, $\theta = t - (z/U)$, where U is the velocity. So, Equation (A1.2), for this special case, is $t = \theta + (z/U)$ and w has to be replaced by θ, x by z, and

$$\begin{aligned}
(\partial z/\partial w)_x &\equiv (\partial z/\partial \theta)_z = 0 \\
(\partial t/\partial w)_x &\equiv (\partial t/\partial \theta)_z = 1 \\
(\partial z/\partial x)_w &\equiv (\partial z/\partial z)_\theta = 1 \\
(\partial t/\partial x)_w &\equiv (\partial t/\partial z)_\theta = 1/U
\end{aligned} \tag{A1.5}$$

Hence, Equation (A1.3) becomes

$$\left(\frac{\partial C}{\partial \theta}\right)_z = \left(\frac{\partial C}{\partial z}\right)_t \left(\frac{\partial z}{\partial \theta}\right)_z + \left(\frac{\partial C}{\partial t}\right)_z \left(\frac{\partial t}{\partial \theta}\right)_z = 0 + 1\left(\frac{\partial C}{\partial t}\right)_z \tag{A1.6}$$

while Equation (A1.4) becomes

$$\left(\frac{\partial C}{\partial z}\right)_\theta = \left(\frac{\partial C}{\partial z}\right)_t \left(\frac{\partial z}{\partial z}\right)_\theta + \left(\frac{\partial C}{\partial t}\right)_z \left(\frac{\partial t}{\partial z}\right)_\theta = \left(\frac{\partial C}{\partial z}\right)_t 1 + \left(\frac{\partial C}{\partial t}\right)_z \left(\frac{1}{U}\right) \tag{A1.7}$$

(The change of boundary conditions is straightforward.)

Example

Rosen [3] started from the equation

$$U\left(\frac{\partial C}{\partial z}\right)_t + \left(\frac{\partial C}{\partial t}\right)_z + \left(\frac{1-\epsilon}{\epsilon}\right)\left(\frac{\partial q}{\partial t}\right)_z = 0$$

C and q being functions of t and z. (Rosen's symbols are used.) The first two terms become $U(\partial C/\partial z)_\theta$ by virtue of Equation (A1.7), while the last term becomes $[(1-\epsilon)/\epsilon](\partial q/\partial \theta)_z$ by virtue of Equation (A1.6). Additionally, Rosen puts $(1-\epsilon)z/\epsilon U = l$. So,

$$\left(\frac{\partial C}{\partial z}\right)_\theta = \left(\frac{\partial C}{\partial l}\right)_\theta \frac{1}{(\partial z/\partial l)_\theta} = \left(\frac{\partial C}{\partial l}\right)_\theta \frac{1-\epsilon}{U\epsilon}$$

Hence, his equation finally became

$$(\partial C/\partial l)_\theta + (\partial q/\partial \theta)_l = 0$$

REFERENCES

[1] J. Edwards, "The Differential Calculus," 3rd. ed., p. 465. Macmillan, New York, 1948.
[2] J. Edwards, "The Differential Calculus," 3rd. ed., Section 158 or 162. Macmillan, New York, 1948.
[3] J. B. Rosen, *J. Chem. Phys.* **20**, 387 (1952).

Appendix 2

Change of Variable in Moving Systems: Distance

Let

$$C = f(z, t) \tag{A2.1}$$

and suppose

$$t = \phi(w, z) \tag{A2.2}$$

as in Appendix 1. If we want to measure distances Z from an origin that is moving with the velocity of the stream, then $Z = z - Ut$; i.e., $z = Z + Ut$. That is to say, w is replaced by Z and x is replaced by t. Hence,

$$(\partial z/\partial w)_x \equiv (\partial z/\partial Z)_t = 1$$

$$(\partial t/\partial w)_x \equiv (\partial t/\partial Z)_t = 0$$
$$(\partial z/\partial x)_w \equiv (\partial z/\partial t)_Z = U$$
$$(\partial t/\partial x)_w \equiv (\partial t/\partial t)_Z = 1$$

Then, as in Appendix 1,

$$\left(\frac{\partial C}{\partial w}\right)_x = \left(\frac{\partial C}{\partial z}\right)_t\left(\frac{\partial z}{\partial w}\right)_x + \left(\frac{\partial C}{\partial t}\right)_z\left(\frac{\partial t}{\partial w}\right)_x$$

which is equivalent to

$$\left(\frac{\partial C}{\partial Z}\right)_t = \left(\frac{\partial C}{\partial z}\right)_t\left(\frac{\partial z}{\partial Z}\right)_t + \left(\frac{\partial C}{\partial t}\right)_z\left(\frac{\partial t}{\partial Z}\right)_t$$

and so becomes

$$\left(\frac{\partial C}{\partial Z}\right)_t = \left(\frac{\partial C}{\partial z}\right)_t(1) + \left(\frac{\partial C}{\partial t}\right)_z 0 = \left(\frac{\partial C}{\partial z}\right)_t$$

by the above relations. Further,

$$(\partial^2 C/\partial Z^2)_t = (\partial^2 C/\partial z^2)_t$$

Again,

$$\left(\frac{\partial C}{\partial x}\right)_w = \left(\frac{\partial C}{\partial z}\right)_t\left(\frac{\partial z}{\partial x}\right)_w + \left(\frac{\partial C}{\partial t}\right)_z\left(\frac{\partial t}{\partial z}\right)_w$$

which is equivalent to

$$\left(\frac{\partial C}{\partial t}\right)_Z = \left(\frac{\partial C}{\partial z}\right)_t\left(\frac{\partial z}{\partial t}\right)_Z + \left(\frac{\partial C}{\partial t}\right)_z\left(\frac{\partial t}{\partial t}\right)_Z$$

and becomes

$$\left(\frac{\partial C}{\partial t}\right)_Z = \left(\frac{\partial C}{\partial z}\right)_t U + \left(\frac{\partial C}{\partial t}\right)_z(1) = U\left(\frac{\partial C}{\partial z}\right)_t + \left(\frac{\partial C}{\partial t}\right)_z$$

So, with the aid of these transformations the equation (for example)

$$D\frac{\partial^2 v}{\partial z^2} - U\frac{\partial v}{\partial z} - \frac{\partial v}{\partial t} = 0$$

becomes

$$D\left(\frac{\partial^2 v}{\partial Z^2}\right)_t = \left(\frac{\partial v}{\partial t}\right)_Z$$

where v is now $v(Z, t)$.

Appendix 3

The Evaluation of Some Integrals

A3.1 PRELIMINARY NOTE

The following identities are used below and are listed for reference:

$$\exp[-(1-\xi)^2/a\xi] = \exp(2/a)\exp[-(1/\xi+\xi)/a] \tag{A3.1}$$

$$\exp \pm[(a^2/\xi^2) + b^2\xi^2] = \exp(\mp 2ab)\exp \pm[(a/\xi) + b\xi]^2 \tag{A3.2}$$

$$= \exp(\pm 2ab)\exp \pm [(a/\xi) - b\xi]^2 \tag{A3.3}$$

$$\int_0^\infty \exp -[(a^2/\xi^2) + b^2\xi^2]\, d\xi \equiv \tfrac{1}{2}I_{-1/2,\infty} = (\pi^{1/2}/2b)\exp(-2ab) \tag{A3.4}$$

(Put $\xi^2 = \zeta$; see Edwards [1].) (For convenience, this appendix adopts the nomenclature

$$I_{n,L} = \int_0^L \xi^n \exp -[(a^2/\xi) + b^2\xi]\, d\xi$$

where L is finite or infinite.)

A3.2 THE INTEGRAL $\int_0^t x^{-3/2} \exp -[(a^2/x) + b^2x]\, dx \equiv I_{-3/2,t}$

This solution is due to Horenstein [2]. The steps in the analysis are

$$\begin{aligned}
\tfrac{1}{2}I_{-3/2,t} &= \int_{1/\sqrt{t}}^{\infty} \exp -[a^2\xi^2 + (b^2/\xi^2)]\, d\xi \qquad (\text{Substitute} \quad \xi = x^{-1/2}) \\
&= \int_{1/\sqrt{t}}^{\infty} f(\xi)\, d\xi = \int_0^{\infty} f(\xi)\, d\xi - \int_0^{1/\sqrt{t}} f(\xi)\, d\xi \\
&= (\pi^{1/2}/2a)(\exp -2ab) - (\exp 2ab) \\
&\quad \times \int_0^{1/\sqrt{t}} \exp -\{a^2[\xi + (b/a\xi)]^2\}\, d\xi
\end{aligned} \tag{A3.5}$$

by Equations (A3.1), (A3.2), and (A3.4); i.e.,

$$\begin{aligned}
\tfrac{1}{2}I_{-3/2,t} &= (\pi^{1/2}/2a)(\exp -2ab) - (b/a)(\exp 2ab) \\
&\quad \times \int_{b\sqrt{t}/a}^{\infty} \lambda^{-2} \exp -\{a^2[(b/a\lambda) + \lambda^2]\}\, d\lambda
\end{aligned}$$

where $\lambda = b/a\xi$; i.e.,

$$\begin{aligned}
\tfrac{1}{2}I_{-3/2,t} &= (\pi^{1/2}/2a)(\exp -2ab) + (\exp 2ab) \int_{b\sqrt{t}/a}^{\infty} [1 - (b/a\lambda^2)] \\
&\quad \times \exp -\{a^2[\lambda + (b/a\lambda)]^2\}\, d\lambda \\
&\quad - (\exp 2ab) \int_{b\sqrt{t}/a}^{\infty} \exp -\{a^2[\lambda + (b/a\lambda)]^2\}\, d\lambda \\
&= (\pi^{1/2}/2a)(\exp -2ab) + (1/a) \\
&\quad \times (\exp 2ab) \int_{b\sqrt{t}+a/\sqrt{t}}^{\infty} (\exp -x^2)\, dx \\
&\quad - (\exp 2ab) \int_{b\sqrt{t}/a}^{\infty} \exp -\{a^2[\lambda + (b/a\lambda)]^2\}\, d\lambda
\end{aligned}$$

where $x = a[\lambda + (b/a\lambda)]$; i.e.,

$$\begin{aligned}
\tfrac{1}{2}I_{-3/2,t} &= (\pi^{1/2}/2a)(\exp -2ab) + (1/a) \\
&\quad \times (\exp 2ab)[(\pi^{1/2}/2) - (\pi^{1/2}/2)\,\mathrm{erf}(bt^{1/2} + at^{-1/2})] \\
&\quad - (\exp 2ab) \int_{b\sqrt{t}/a}^{\infty} \exp - \{a^2[\lambda + (b/a\lambda)]^2\}\, d\lambda
\end{aligned}$$

That is to say,

$$\tfrac{1}{2}I_{-3/2,t} = (\pi^{1/2}/a)(\cosh 2ab) - (\pi^{1/2}/2a) \times (\exp 2ab)[\operatorname{erf}(bt^{1/2} + at^{-1/2})] - (\exp 2ab)\int_{b\sqrt{t}/a}^{\infty} \exp -\{a^2[\lambda + (b/a\lambda)]^2\}\, d\lambda \quad \text{(A3.6)}$$

Again, if instead of Equation (A3.2), Equation (A3.3) is used in the development from Equation (A3.5) onwards, followed by the same substitutions, the result is

$$\tfrac{1}{2}I_{-3/2,t} = (\pi^{1/2}/2a)(\exp -2ab)[\operatorname{erf}(bt^{1/2} - at^{-1/2})] + (\exp -2ab)\int_{b\sqrt{t}/a}^{\infty} \exp -\{a^2[\lambda - (b/a\lambda)]^2\}\, d\lambda \quad \text{(A3.7)}$$

Hence, the sum of Equations (A3.6) and (A3.7) [and the use of identities (A3.2) and (A3.3)] will give

$$I_{-3/2,t} \equiv \int_0^t x^{-3/2} \exp -[(a^2/x) + b^2x]\, dx = (\pi^{1/2}/a)(\cosh 2ab) + (\pi^{1/2}/2a)f(t) \quad \text{(A3.8)}$$

where

$$f(t) = (\exp -2ab)[\operatorname{erf}(bt^{1/2} - at^{-1/2})] - (\exp 2ab)[\operatorname{erf}(bt^{1/2} + at^{-1/2})]$$

A3.3 THE INTEGRAL $I_{-1/2,t}$

Since

$$I_{-1/2,t} \equiv \int_0^t x^{-1/2} \exp -[(a^2/x) + b^2x]\, dx = (-1/2b)\, d(I_{-3/2,t})/db$$

it follows that differentiation of (A3.8) gives

$$I_{-1/2,t} = (\pi^{1/2}/2b)[f(t) - \sinh 2ab] \quad \text{(A3.9)}$$

where

$$f(t) = (\exp -2ab)[\operatorname{erf}(bt^{1/2} - at^{-1/2})] + (\exp 2ab)\operatorname{erf}(bt^{1/2} + at^{-1/2})$$

The above derivation of Equation (A3.9) is, for the most part, straightforward, but the following note may be helpful. Since $(d/db)\int_0^b y\, dx = y(b)$, so

$$\frac{d}{db}\int_0^{f(b)} y\, dx = \frac{df(b)}{db}\frac{d}{df(b)}\int_0^{f(b)} y\, dx = y[f(b)]\frac{df(b)}{db}$$

Thus,

$$(d/db)[\operatorname{erf}(bt^{1/2} \pm at^{-1/2})] = [2t^{1/2}\pi^{-1/2}] \exp[-(bt^{1/2} \pm at^{-1/2})^2]$$

A3.4 THE INFINITE INTEGRALS

By taking t to the limit $t \longrightarrow \infty$ there results

$$I_{-3/2,\infty} = (\pi^{1/2}/a) \exp -2ab \tag{A3.10}$$

$$I_{-1/2,\infty} = (\pi^{1/2}/b) \exp -2ab \qquad \text{(A3.4) R}$$

A3.5 OTHER INTEGRALS OF INTEREST

Other integrals that may arise in this subject are

$$I_{-1,\infty} = \int_0^\infty \xi^{-1} \exp -(a^2\xi^{-1} + b^2\xi)\, d\xi = 2K_0(2ab) \tag{A3.11}$$

where K_0 denotes a modified Bessel function of the second kind. (See, for example, Gröbner and Hofreiter [3] or Wilson [4].)

$$I_{1/2,\infty} = [(1 + 2ab)/2b^3]\pi^{1/2} \exp -2ab \tag{A3.12}$$

(Gröbner and Hofreiter [3]);

$$I_{3/2,\infty} = [(3 + 6ab + 4a^2b^2)/4b^5]\pi^{1/2} \exp -2ab \tag{A3.13}$$

(Gröbner and Hofreiter [3]).

A3.6 SOME INTEGRALS ARISING FROM TAKING MOMENTS OF A DISTRIBUTION

The infinite integrals having the general form

$$I = \int_0^\infty t^n \exp[-(z - Ut)^2/4Dt]\, dt$$

(which arise when computing moments of the concentration distribution (with time) caused by a perfect impulse in one-dimensional flow) can quickly be transformed into

$$I = (z/U)^{n+1}[\exp(\text{Pe}/2)] \int_0^\infty \theta^n \exp[-(\text{Pe}/4\theta) + (\text{Pe}\theta/4)]\, d\theta$$

$$= (z/U)^{n+1}[\exp(\text{Pe}/2)](I_{n,\infty})$$

In the latter integral $a = b = \text{Pe}^{1/2}/2$, and so it quickly follows that for $n = -\frac{1}{2}$, by Equation (A3.4)

$$I = 2(z/U)^{1/2}(\pi/\text{Pe})^{1/2} \tag{A3.14}$$

while for $n = \frac{1}{2}$, by Equation (A3.12)

$$I = 4(z/U)^{3/2}\pi^{1/2}\{1 + [(\mathrm{Pe}/2)/\mathrm{Pe}^{3/2}]\} \tag{A3.15}$$

If (still for $n = \frac{1}{2}$), $U = 0$, the definite integral

$$I = \int_0^z t^{-1/2} \exp(-\xi^2/4Dt)\, d\xi \equiv 2D^{1/2} \int_0^{z/2\sqrt{(Dt)}} \exp(-\zeta^2)\, d\zeta$$

[substitute $\zeta = \xi/(2D^{1/2}t^{1/2})$]

$$= D^{1/2}\pi^{1/2} \operatorname{erf}[z/(2D^{1/2}t^{1/2})] \tag{A3.16}$$

Finally, for $n = \frac{3}{2}$, by Equation (A3.13)

$$I = 8(z/U)^{5/2}\pi^{1/2}[3 + (3\mathrm{Pe}/2) + (\mathrm{Pe}^2/4)]/\mathrm{Pe}^{5/2} \tag{A3.17}$$

The dimensions of the answers in Equations (A3.14)–(A3.17) are $[\mathrm{T}]^{1/2}$, $[\mathrm{T}]^{3/2}$, $[\mathrm{L}][\mathrm{T}]^{-1/2}$, and $[\mathrm{T}]^{5/2}$, as they should be from the definition of the integrals.

REFERENCES

[1] J. Edwards, "Integral Calculus." Chelsea, New York, 1954.
[2] W. Horenstein, *Quart. Appl. Math.* **3**, 183 (1945).
[3] W. Gröbner and N. Hofreiter, "Integraltafel." Springer-Verlag, Berlin, 1961.
[4] H. A. Wilson, *Proc. Cambridge Phil. Soc.* **12**, 406 (1904).

Appendix 4

Some Properties of the Equation $y = y_0 \exp(-k^2 z^2)$ and of Its Integral

A4.1 THE FINDING OF THE VARIANCE OF THE CURVE

The curve of $y = y_0 \exp(-k^2 z^2)$ is shown in Figure A4.1 as curve A, and that of

$$I = y_0 \int_0^z \exp(-k^2 \xi^2)\, d\xi = (y_0/k) \int_0^{kz} \exp(-\zeta^2)\, d\zeta$$

as curve B.

Curve A has the general shape of the normal distribution curve. Considered as a frequency distribution, it would have a variance

$$\sigma^2 = \int_{-\infty}^{\infty} \xi^2 y \, d\xi \Big/ \int_{-\infty}^{\infty} y \, d\xi = 2y_0 \int_0^{\infty} \xi^2 \exp -(k\xi)^2 \, d\xi \Big/ 2y_0 \int_0^{\infty} \exp -(k\xi)^2 \, d\xi$$

$$= (2y_0/2k^3) \int_0^{\infty} z^{1/2} \exp -z \, dz \Big/ (2y_0/k) \int_0^{\infty} \exp -z^2 \, dz$$

[substitute $z = (k\xi)^2$ for the numerator and $z = k\xi$ for the denominator]

$$= (1/2k^2)\Gamma(3/2)/[(\pi^{1/2}/2) \operatorname{erf} \infty] = (1/2k^2\pi^{1/2})\Gamma(1/2) = 1/2k^2$$

Hence,

$$\sigma^2 = 1/2k^2 \tag{A4.1}$$

The points y_a, I_a, and I_b on the appropriate curves have ordinates fixed by the criteria developed below and the numerical values of the corresponding abscissas, measured from the line of symmetry of curve A, will then lead to values of the standard deviation of the curve A (in the units of the abscissas). For, if $y_a = y_0 \exp -1 = y_0 \exp - (kz_a)^2$, then by Equation (A4.1)

$$z_a = 1/k = 2^{1/2}\sigma \tag{A4.2}$$

Also,

$$I_a/I_\infty = \int_0^{kz_a} \exp -\zeta^2 \, d\zeta \Big/ \int_0^{\infty} \exp -\zeta^2 \, d\zeta$$

$$= (\operatorname{erf} kz_a)/\operatorname{erf} \infty = \operatorname{erf} 1/\operatorname{erf} \infty$$

and again $z_a = 2^{1/2}\sigma$, by Equation (A4.2). Hence,

$$I_a/I_\infty = 0.8427 \quad \text{at} \quad z_a = 2^{1/2}\sigma \tag{A4.3}$$

Similarly, at $z_b = \sigma$ [where $kz_b = k\sigma = 2^{-1/2}$ by Equation (A4.1)] the ordinate I_b is such that

$$I_b/I_\infty = (\operatorname{erf} kz_b)/\operatorname{erf} \infty = \operatorname{erf}(2^{-1/2}) = 0.6827 \tag{A4.4}$$

and $y_b/y_0 = e^{-1/2} = 0.6065 = 0.6827 \times \pi^{1/2}/2$. Since $I(\infty) = (\pi^{1/2}/2)y_0$, it follows that the two curves intersect at this point.

The use of Equation (A4.4) would probably be preferable to that of Equation (A4.2) or (A4.3) because (a) at the larger values of z, experimental inaccuracies are likely to be greater; (b) both curves A and B show a relatively small rate of change of ordinate per unit change of abscissa in this region; and (c) failure of the diffusion model shows up more at the extremes.

If a step function is used, it is not necessary to find the midheight of the resulting concentration–time curve (i.e., the origin of the curve B in Figure A4.1); in practice the method of the example in Section 3.6.2 would be used. The values given in Figure 3.14 arise directly from Equation (A4.4) and Figure A4.1. For example, θ_b is at $v = \frac{1}{2}(1 + 0.6827) \doteqdot 0.84$.

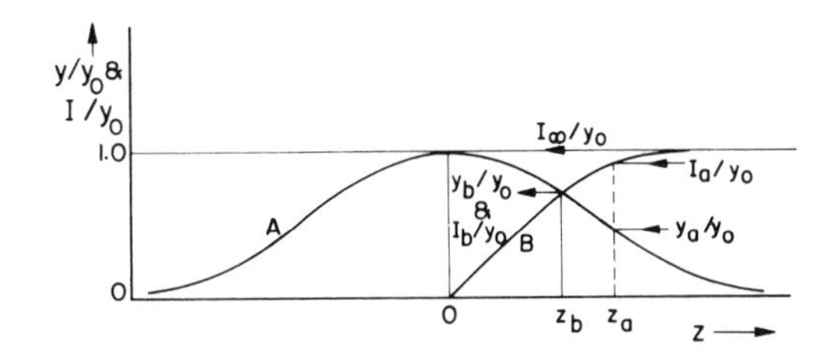

Fig. A4.1. Curves of $y = y_0 \exp(-k^2z^2)$ and of its integral.

A4.2 THE CONDITIONS FOR NORMALIZATION

The curve of y is said to be normalized if the area under the curve is equal to unity; to do this, the values of the parameters must be chosen to have a value that is calculated thus:

$$\text{Area} = 1 = 2y_0 \int_0^\infty \exp -(k\xi)^2 \, d\xi \equiv (2y_0/k) \int_0^\infty \exp -(\zeta)^2 \, d\zeta$$

(substitute $\zeta = k\xi$)

$$= 2y_0\pi^{1/2}/2k$$

by Section A4.1. Hence, $y_0 = k\pi^{-1/2} = 1/(2^{1/2}\pi^{1/2}\sigma)$ by Equation (A4.1), and so the equation to the normal curve of unit area and of variance σ^2 is given by

$$y = [1/(2\pi)^{1/2}\sigma] \exp[-(z^2/\sigma^2)] \qquad \text{(A4.5)}$$

Appendix 5

The Development of a Laplace Transform of a Pulse

In Section 2.3.4(b) there arose

$$\bar{f}_r(s+a) \equiv \bar{v}_r = (q_{\mathrm{pl}}/2D)\{1/s[(a/D)+(s/D)]^{1/2}\} \times \exp\{(U/2D) \pm [(a/D)+(s/D)]^{1/2}\}z$$

where $r = 1, 2$, $a = U^2/4D$, and the positive sign relates to $r = 1$. Rearranged, this is

$$\bar{f}_r(s+a) = (q_{\mathrm{pl}}/D)[D^{1/2}/s(a+s)^{1/2}] \exp[\pm(a+s)^{1/2}zD^{-1/2}] \exp(Uz/2D) \tag{A5.1}$$

The table of transforms gives the inverse transform of

$$(D/s)^{1/2} \exp - (s^{1/2}z/D^{1/2})$$

as $(D/\pi t)^{1/2} \exp(-z^2/4Dt)$, and since the inverse transform of $\bar{f}(s + \mathrm{a})$ is $[\exp - (at)]f(t)$ and since that of $(1/s)\bar{f}(s) = \int_0^t f(t)\, dt$, it follows that for *region* 2

$$\begin{aligned} v_2 &= \text{inverse transform of } \bar{f}_2(s + a) \\ &= (q_{\mathrm{pl}}/2D^{1/2}\pi^{1/2})[\exp(Uz/2D)] \int_0^t t^{-1/2} \exp -[(U^2t/4D) + (z^2/4Dt)]\, dt \end{aligned} \tag{A5.2}$$

z being positive. It can be written as

$$v_2 = \{q_{\mathrm{pl}}[\exp(Uz/2D)]/2(\pi D)^{1/2}\}I_{-1/2,t}$$

(Appendix 3, in which $a = z^2/4D$, $b = U^2/4D$).

For *region* 1, z is negative, so Equation (A5.1) becomes

$$\begin{aligned} \bar{f}_1(s + a) = {} & (q_{\mathrm{pl}}/D)\{\exp[-U(-z)/2D]\} \\ & \times \{[D^{1/2}/s(a + s)^{1/2}] \exp[-(a + s)^{1/2}D^{-1/2}(-z)]\} \end{aligned}$$

and so

$$\begin{aligned} v_1 &= f(t) \\ &= [q_{\mathrm{pl}}/(\pi D)^{1/2}]\{\exp[-U(-z)/2D]\} \\ &\quad \times \int_0^t t^{-1/2} \exp -\{(U^2t/4D) + [(-z)^2/4Dt]\}\, dt \\ &= \{[q_{\mathrm{pl}}/(\pi D)^{1/2}] \exp[-U(-z)/2D]\}I_{-1/2,t} \end{aligned} \tag{A5.3}$$

where $a = (-z)^2/4D$, $b = U^2/4D$, and $(-z)$ is positive; i.e., measurements are taken as positive in the upstream direction.

When $t \longrightarrow \infty$ the concentration acquires the steady-state distribution, Equation (2.20), as it should. For $I_{-1/2,\infty} = (\pi^{1/2}/b) \exp -2ab$, see Equation (A3.4). Then, for *region* 2, substitution for $a = z^2/4D$ and $b = U^2/4D$ into the last equation changes Equation (A5.2) into

$$(v_2)_{\mathrm{steady}} = [q_{\mathrm{pl}}/2(\pi D)^{1/2}]e^{Uz/2D}[2(\pi D)^{1/2}/U]e^{-Uz/2D} = q_{\mathrm{pl}}/U$$

as Equation (2.20a).

For *region* 1, substitution of $a = (-z)^2/4D$ and $b = U^2/4D$ similarly changes Equation (A5.3) into

$$(v_1)_{\mathrm{steady}} = [q_{\mathrm{pl}}/2(\pi D)^{1/2}]e^{-U(-z)/2D}[2(\pi D)^{1/2}/U]e^{-U(-z)/2D} = (q_{\mathrm{pl}}/U)e^{-U(-z)/D}$$

as Equation (2.20b), $(-z)$ being positive.

Appendix 6

The Laplace Transform of Equation (2.36)

The condition given by Equation (2.36) was

$$Q_{\mathrm{pl}} = \int_{-\infty}^{\infty} v\,dZ = 2\int_{0}^{\infty} v\,dZ$$

Hence, a Laplace transformation gives

$$Q_{\mathrm{pl}}/s = 2\int_{0}^{\infty}\left(\int_{0}^{\infty} v\,dZ\right)e^{-st}\,dt = 2\int_{0}^{\infty}\left(\int_{0}^{\infty} v\,e^{-st}\,dt\right)dZ = 2\int_{0}^{\infty} \bar{v}\,dZ$$

Appendix 7

The Variance of a Distribution

(Aitken [1] contains clear expositions: probability function [1, Chapter 1, p. 16]; moments [1, Chapter 1, p. 9]; probability and distribution curves, the parameters associated with them, and their interrelations [1, Chapter 2].)

Consider a collection of numerical quantities having values designated by $X_1, X_2, \ldots, X_r, \ldots, X_n$, there being N_r quantities of value X_r, while $\sum^n N_r = N$. Their arithmetic mean is designated by $\bar{X}$, defined by

$$\sum^n N_r X_r / N\dagger$$

†For small collections there are reservations about what the mean actually is, but these are ignored here.

If the distribution is continuous, then the number of terms N_r of value X_r is now to be replaced by the amount of the variable whose value lies in the small range $X - \frac{1}{2}\,dX$ and $X + \frac{1}{2}\,dX$; this amount of the variable is linked to the width of the interval dX. It is denoted by $f'(X)\,dX$; i.e., it can be thought of as an area of height $f'(x)$ and width dX. The total area $= \int_{-\infty}^{\infty} f'(X)\,dX$ $=$ total amount A of the variable. Alternatively, the relative amount can be used: $f(x) = f'(X)/A$. So, $\int_{-\infty}^{\infty} f(X)\,dX = 1$, and the distribution is thus normalized. The term $f(X)$ can be used to apply to both discrete and continuous distributions. For both, the rth moment of a frequency distribution is *defined* as $m_r^* = \sum^n X^r f(X)$ or $\int_{-\infty}^{\infty} X^r f(X)\,dX$. The first moment will thus be the mean of the distribution, already referred to. If measurements are to be taken from this mean, then $X - \bar{X} = x$, and so $m_r = \sum^n x^r f(x)$ or $\int_{-\infty}^{\infty} x^r f(x)\,dx$. The moment m_r can be expressed in terms of m_r^* together with other "asterisked" moments by expanding the appropriate definition by the binomial expansion. In particular,

$$m_0 = 1$$

$$m_1 = 0$$

$$m_1^* = \bar{X} = \sum^n Xf(X) \qquad \text{or} \qquad \int_{-\infty}^{\infty} Xf(X)\,dX$$

$$\sigma^2 = m_2 = \sum^n (X - \bar{X})^2 f(X) \qquad \text{or} \qquad \int_{-\infty}^{\infty} (X - \bar{X})^2 f(X)\,dX$$

[*N.B.*: $f(X)$ can be written as $f(x)$, of course]; it follows that

$$\sigma^2 = \sum^n (X^2 + \bar{X}^2 - 2X\bar{X}) f(x)$$

or

$$\sigma^2 = \int_{-\infty}^{\infty} (X^2 + \bar{X}^2 - 2X\bar{X}) f(x)\,dx$$

Hence,

$$\sigma^2 = \sum^n X^2 f(x) + \bar{X}^2 \sum^n f(x) - 2\bar{X} \sum^n Xf(x)$$

or

$$\sigma^2 = \int_{-\infty}^{\infty} X^2 f(x)\,dx + \bar{X}^2 \int_{-\infty}^{\infty} f(x)\,dx - 2\bar{X} \int_{-\infty}^{\infty} Xf(x)\,dx$$

So

$$\sigma^2 = m_2^* + \bar{X}^2 - 2\bar{X}^2$$

in both cases since $\sum^n f(x)$ and $\int_{-\infty}^{\infty} f(x)\,dx$ equal unity. Hence, generally

$$\sigma^2 = m_2^* - (m_1^*)^2$$

The mean can have any numerical value, depending on the datum used (for example, the mean time measured from a temperature–time curve will depend upon the datum used). The higher moments likewise have values that depend upon the origin of X. The unstarred moments, having x measured from the mean, have unique values for any given distribution.

If the concentration–time curve is thought of as a probability curve rather than a frequency distribution curve, then the argument still holds, but $f(x)$ is usually replaced by $\phi(x)$, while μ is used in place of m: see Aitken [1].

REFERENCE

[1] A. C. Aitken, "Statistical Mathematics." Oliver & Boyd, Edinburgh, 1947.

Appendix 8

Transport Equations

The Use of Concentrations of Matter or Heat

Equation (1.5a) is the one governing the systems through which plane kinematic waves travel. It may be called a *wave equation of unreciprocal dispersive systems*, and as such it can be derived by combining the appropriate *first and second telegrapher's equations*.

It is also known as the *dispersion model equation*, and as such it can be derived from a differential balance over length interval dz ($\longrightarrow 0$) in time interval dt ($\longrightarrow 0$); texts on transport phenomena contain the derivation. Fulford and Pei [1] contains a full discussion. There have to be assumptions made about the space-averaged and time-averaged values of the variables. For both mass and heat transfer equations the basic equation is a balance of mass or heat, but the latter is almost invariably converted into temperatures.

One reason is probably that temperature seems to be more readily measurable and more in line with everyday experience than "concentration of heat"; another is that the transfer equations for matter and heat have traditionally been written differently. Thus, they are as follows, where $\tilde{q}$ is a flux of matter or heat per unit area, and with the sign convention of Chapter 7.

Mass Transport

$$-\tilde{q} = k\{v_1 - [(v_j)_{\text{surf}}/\mathbf{K}]\} \tag{A8.1}$$

where k is a mass transfer coefficient and $(v_j)_{\text{surf}}/\mathbf{K}$ gives the concentration in the fluid in thermodynamic equilibrium with the concentration on the surface of the jth phase. (The assumption is often made that $\mathbf{K}$ is constant; if so, then the thermodynamic conditions must be that the temperature is uniform and constant, and that Henry's law holds, relating fugacity to concentration in all phases linearly.)

Heat Transport

$$-\tilde{q} = h[T_1 - (T_j)_{\text{surf}}]$$

where h is a heat transfer coefficient and T is temperature.

If, however, this is written in terms of concentration of heat (i.e., $v = FT$), then

$$-\tilde{q} = k\{v_1 - [(v_j)_{\text{surf}}/\mathbf{K}]\} \tag{A8.2}$$

where $k = h/F_1$; $v_1 = F_1T_1$; $(v_j)_{\text{surf}} = F_j(T_j)_{\text{surf}}$, and $\mathbf{K} = F_j/F_1$. If F_j and F_1 are constant, then $\mathbf{K}$ is also constant; the conditions are therefore that composition and temperature do not change very much. Further, if the pressure is constant, then $F = [\partial(\rho H)/\partial T]_p$, where ρ is density and H the specific enthalpy.

Thus, Equations (A8.1) and (A8.2) are of the same form, and for both, appropriate conditions are attached. In both, $\tilde{q}$ is amount/(unit time)(unit area); k has dimensions $[L][T]^{-1}$ and v is amount/(unit volume).

(The Nomenclature Section also lists k'; k_1 and k_j when the "forward" and "backward" rates are different, as well as κ.)

The relation between the flux per unit volume of flowing phase, q, is related, via the shunt admittance, to the "driving force" in the flowing phase. The latter may be concentration, as here, but the admittance may have been derived for thermal systems where temperatures were used. There is a simple relation between the two admittances, for, from Equation (7.3a)

$$q_j = -\tilde{V}Y_C v_1 \qquad \text{and} \qquad q_j = -\tilde{V}Y_T T_1$$

where Y_C and Y_T are the appropriate admittances in the respective cases, and T_1 is the instantaneous temperature in phase 1. Since $v_1 = F_1 T_1$, then with appropriate units

$$Y_C = Y_T / F_1$$

Definition of Resistance R

The resistance R is dimensionally like impedance Z. Hence, since $Z = 1/Y$, and $Y = (\tilde{V} \times \text{"driving force"})/q$, while the driving force $= \{v_1 - [(v_j)_{\mathrm{surf}}/\mathbf{K}]\}$, and $q_j = \tilde{a}k\{v_1 - [(v_j)_{\mathrm{surf}}/\mathbf{K}]\}$, then $R = \tilde{V}/\tilde{a}k$.

REFERENCE

[1] G. D. Fulford and D. C. T. Pei, A Unified Approach to the Study of Transfer Processes. *Ind. Eng. Chem.* **61**(5), 47 (1969).

Appendix 9

To Show That $(b \pm ig)^{1/2} = [(s + b)/2]^{1/2} \pm i[(s - b)/2]^{1/2}$

Let

$$b = s \cos \alpha, \qquad g = s \sin \alpha$$

where α is a variable. Then, $s = (b^2 + g^2)^{1/2}$, positive root; and $\cos \alpha = b/s = b/(b^2 + g^2)$. So,

$$b \pm ig = s(\cos \alpha \pm i \sin \alpha) = se^{\pm i\alpha}$$

Hence,

$$(b \pm ig)^{1/2} = s^{1/2}e^{\pm i\alpha/2} = s^{1/2}[\cos(\alpha/2) \pm i \sin(\alpha/2)]$$

But

$$\cos(\alpha/2) = [(1 + \cos \alpha)/2]^{1/2} = [(1 + b/s)/2]^{1/2}$$

and

$$\sin(\alpha/2) = [1 - \sin(\alpha/2)]^{1/2} = [(1 - b/s)/2]^{1/2}$$

Hence,

$$\begin{aligned} b \pm ig &= s^{1/2}\{[(1 + b/s)/2]^{1/2} \pm i[(1 - b/s)/2]^{1/2}\} \\ &= [(s + b)/2]^{1/2} \pm i[(s - b)/2]^{1/2} \end{aligned}$$

Appendix 10

General Wave Properties. Interference

A number of the attributes and properties of waves—e.g., frequency, velocity, wavelength, attenuation, impedance, and normal reflection—have been described in this book. It would be possible to discuss some others—e.g., diffraction, refraction, scattering, coherence, oblique reflection, and interference—that kinematic waves can also show. The last-mentioned phenomena are explained in texts on waves; for example, Sharma [1], or, in more detail, Rayleigh [2], and some, at least, can be readily related to behavior discussed in this book.

To one using kinematic waves for the determination of parameters, only the phenomenon of interference appears to be of possible interest, and this particularly in the case where an incident wave divides into a number of paths. Chapter 9 gives an example. Differences of path length, fluid velocity, and dynamic characteristics of each path will cause the amplitude and phase angle of the wave issuing from each to be different. The (averaged) wave as measured at the outlet will be the vector sum of all these separate waves, as discussed in, say, Sharman [1], but with the difference that the wave from each path will be weighted by the flow rate of fluid in that path. The question of reflections is not considered. These, in any case, will not arise if "impedance matching" holds; in practice this would make it necessary to assume that all streams are brought together (with no change of impedance occurring) then passing immediately into a perfect mixer of vanishingly small size just downstream of the confluence. This mixer must have characteristics such that, again, no impedance change is "seen" by the entering stream.

Figure A10.1(a) shows a collection of such paths (each of which could be, for example, a channel in a packed bed or porous rock, or capillaries or tracheae in living organisms). A plane wave front at A may, by Huyghen's principle, be divided up into plane wavefronts in each of the channels. In the infinitesimal mixing zone at B a detector averages the wave across the

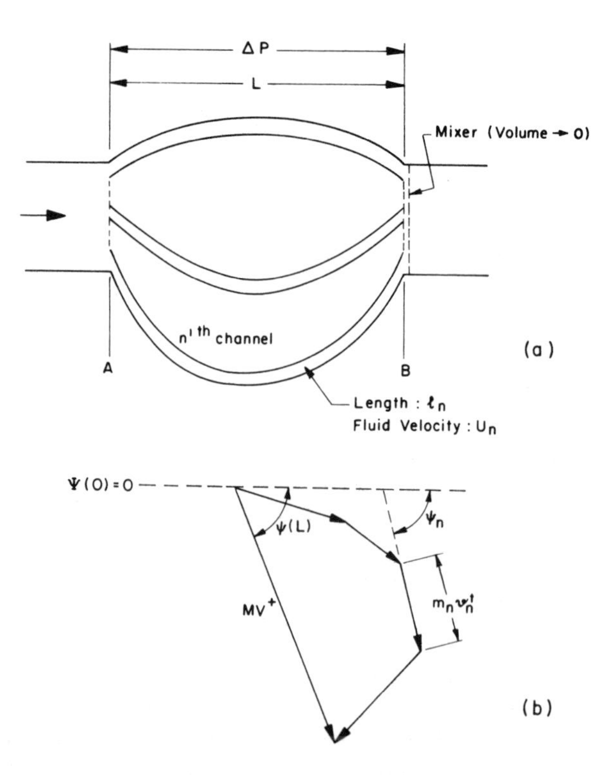

Fig. A10.1. Interference of kinematic waves.

outlet duct. If v_n^+ is the amplitude, ψ_n the phase angle (relative to the inlet wave), and m_n the flow rate, all in the nth channel, then the measured positive issuing wave is $\mathbf{V}^+e^{i[\omega t-\psi(L)]}$, where

$$M\mathbf{V}^+ = [(\sum^n m_n v_n^+ \cos \psi_n)^2 + (\sum^n m_n v_n^+ \sin \psi_n)^2]^{1/2}$$

$$\tan \psi(L) = (\sum^n m_n v_n^+ \sin \psi_n)/(\sum^n m_n v_n^+ \cos \psi_n)$$

and

$$M = \sum^n m_n = \text{total flow rate}$$

In the above, v_n^+ and ψ_n will be functions of ω, l_n, and U_n, and of the dynamic characteristics of the channel, while m_n will be a function of ΔP (the pressure drop), the hydraulic characteristics of the channel, the flow regime, and the characteristics of the fluid. The net wave, as observed in a system of nominal length L will have a "velocity" $V_w = L\omega/\psi$ and a "wavelength" $\lambda = 2\pi L/\psi$, where ψ is the observed phase lag referred to in Equations (9.24) and (9.25).

Figure A10.1(b) shows the vector polygon. It is possible for the closing $M\mathbf{V}^+$ to be zero, i.e., for $\mathbf{V}^+$ to be zero. Thus, at certain frequencies there would be no wave issuing from the bed, the interference of all the component waves being such that they neutralize each other.

This concept of interference can be applied to flow in a single channel. For example, when there is laminar flow with no lateral diffusion, the wave in each filament of fluid is unaffected by its neighbor, and the net effect (as detected) may be obtained by using the equations in this appendix. Thus, based on Kramers and Alberda [3], Poiseuille flow with a wave $V_S e^{i\omega t}$ at $z = 0$ will, under the above conditions, give rise to a positive wave at point z of

$$v^+ \cos \psi = V_S[\cos N_f - N_f \sin N_f + N_f^2 \mathrm{Ci}(N_f)]$$

and

$$v^+ \sin \psi = V_S\{\sin N_f + N_f \cos N_f - N_f^2[\pi/2 - \mathrm{Si}(N_f)]\}$$

where

$$N_f \equiv \text{frequency number} = \omega z/U_{\max},$$

$$\mathrm{Si}(N_f) \equiv \text{sine integral} = \int_0^{N_f} (\sin \xi/\xi)\, d\xi,$$

$$\mathrm{Ci}(N_f) \equiv \text{cosine integral} = -\int_{N_f}^{\infty} (\cos \xi/\xi)\, d\xi,$$

$$U_{\max} = \text{maximum velocity (at } r = 0) \text{ in steady Poiseuille flow}$$

while the mass flow rate in an annulus of mean radius r and thickness dr (which will be needed in the derivation) will be

$$dm(r) = 2\pi \rho r U(r)\, dr$$

and

$$U(r) = U_{max}[1 - (r/r_0)^2]$$

in a capillary of radius r_0. [Note that the exponential expression of Equation (6.10) does not apply in this case.]

For the behavior of an impulse or step-change in Poiseuille flow, Taylor [4] may be consulted. When there is Poiseuille flow and diffusion is also present, then the independence of each filament or stream tube is violated; in fact, each acts as a reservoir phase to its neighbors. The behavior of sine waves in such a system (with isotropic diffusion) has been covered by Carrier [5].

REFERENCES

[1] R. V. Sharman, "Vibrations and Waves." Butterworth, London and Washington, D.C., 1963.
[2] Lord Rayleigh, "The Theory of Sound," 2 vols. Dover, New York, 1945.
[3] H. Kramers and G. Alberda, Frequency Response Analysis of Continuous Flow Systems. *Chem. Eng. Sci.* **2**, 173 (1953).
[4] G. I. Taylor, Conditions under which Dispersion of a Solute in a Stream of Solvent can be used to Measure Molecular Diffusion. *Proc. Roy. Soc. (London)* **A225**, 473 (1954).
[5] G. F. Carrier, On Diffusive Convection in Tubes. *Quart. J. Appl. Math.* **14**, 108 (1956).

Author Index

Numbers in parentheses are reference numbers and indicate that an author's work is referred to, although his name is not cited in the text. Numbers in italics show the page on which the complete reference is listed.

Subject Index

T